Piggery Waste Management

Piggery Waste Management

Towards a Sustainable Future

Euiso Choi

IWA Publishing

Published by IWA Publishing, Alliance House, 12 Caxton Street, London SW1H 0QS, UK
Telephone: +44 (0) 20 7654 5500; Fax: +44 (0) 20 7654 5555; Email: publications@iwap.co.uk
Web: www.iwapublishing.com

First published 2007

British Library Cataloguing in Publication Data
A CIP catalogue record for this book is available from the British Library

Library of Congress Cataloging- in-Publication Data
A catalog record for this book is available from the Library of Congress

ISBN: 1843391317
ISBN13: 9781843391319

Contents

Preface

The Chinese calendar says the year of 2007 is the year of the pig, which they believe brings wealth, because a pig delivers many baby pigs to get money and many pigs produce more waste to apply to an agricultural area as fertilizer. Eventually, China has become the country growing the largest number of pigs in the world. The number of pigs and the sizes of pig farms have been growing to meet the increased demand of meat, while the land area for agriculture has been reduced. The result is over saturation of nutrients in soil, and the piggery waste problem has become an international problem.

Korea is a small country with a large population. The Koreans love to eat bacon and grow many pigs resulting in a severe pollution problem. Therefore, the Korean government enforced stringent regulations on animal waste management. This has become a motivation to develop new technologies for piggery waste management with limited land.

The author started research on nightsoil which is a technical word for human excreta in earlier days. It seems that the technology applicable to nightsoil is similar to that used for municipal sludge, but there is a big difference. The major purpose of municipal sludge treatment is a volumetric reduction, but the purpose of nightsoil treatment is a stabilization of organic substances. The piggery waste has been a simple problem when the whole waste is applied to land. But the problem becomes complicated in a similar way to that of nightsoil when it needs to be treated. In general, most people including environmental engineers are confused over what kinds of technologies should be applied to piggery waste management towards a sustainable future. Environmental engineers have a tendency to directly apply sludge treatment technology for piggery waste management. It may be a wrong approach. It is the purpose of this book to enable readers to understand the problem correctly and to lead them towards the right approach to the proper management of piggery waste, which has become an international problem both in developed and developing countries for a sustainable future.

The author would like to express his gratitude to his colleagues who have provided technical assistance, and to his students who have provided drafting work. Also, sincere thanks must be extended to his family for their kind patience, particularly for his lovely daughter who has provided the wonderful art work for the cover of this book.

Hope you have a wonderful dream of wealth with this book from the year of pig.

May, 2007

Dr. Euiso Choi
Professor Emeritus of Civil and Environmental Engineering,
Korea University, Seoul,
Korea.

1

Introduction

This book is about piggery waste management. Piggery wastes contain high concentrations of organics and nitrogen. Other environmental problems associated with piggery wastes include heavy metals (zinc and copper) (Bull, 2005) and antibiotics (Campagnolo et al., 2002). Besides being a water pollution problem, agriculture is a major contributor to green house gas (GHG) emissions from three major sources: manure, which can result in the uncontrolled release of methane; enteric fermentations of ruminant animals; and release of N_2O from the soil as a result of fertilizer application. These sources contribute approximately 10% of GHG emissions in Europe, with manure management giving rise to an estimated 1.7 million tonnes of CO_2 equivalent (Banks et al., 2006).

There are many different names for the word *pig* in English and Chinese as shown in Figure 1.1. Pigs have been domesticated since human history began. There are still wild pigs, but mostly tamed with other animals such as dogs, cats and cows. To this day neither an orthodox Jew nor a Muslim will eat pork, but Christian communities are normally allowed to eat pigs.

The term *hog* has often been used for full-grown domestic swine. The term *pig* is generally applied to younger or smaller hogs, although it is also often used for domestic swine in general.

豚 猪 亥

Swine, pig, hog, boar, sow, barrow, gilt

Figure 1.1 Different names of pig in English and Chinese.

Pigs were among the first animals to be tamed by men as early as 2900 BC. Tales of wild-boar hunts have come down from the Middle Ages. Boar's heads often decorated the banquet tables of those days. Hog breeds have in the past been conveniently divided into two types: bacon and lard. Lard hogs have been most numerous and popular in the U.S. They are fat-backed, broad, deep animals with much more fat on them than lean meat. In recent years there has been a tendency to change the lard type into what is called a "meat type" hog somewhat intermediate between the old lard and bacon types. The Poland China is a large breed whose ancestors were a combination of Byfield, Big China, Russian, Irish Grazier, and Berkshire breeding. The males, called boars, may weigh as much as 1,000 pounds (about 500 kg); and the females, called sows, as much as 800 pounds (400 kg). Hogs furnish some of the most appetizing and most commonly eaten meat dishes. These include bacon, ham, pork chops and roasts, pig's knuckles, sausage, and pig's jowls. Their fat yields lard, which is almost a necessity in the kitchen. Their hides give excellent leather pigskin. Their hair is used as stuffing for mattresses. Some of the best toilet brushes are made from their bristles. Glue is made from their hoofs; fertilizer, from their bones and sausage casings, from their intestines (Britannica Junior Encyclopedia, 1979).

Hogs do not like dirty places. They do not have a well controlled body temperature system by sweating moisture through skin, but like to cool their body by submerging themselves into water. Thus, they drink water and urinate to control their body temperature. Hogs noses are well developed and they know where they can discharge their feces and urine in order to keep their places clean.

Hogs breed rapidly. Often there are as many as 12 pigs in a litter, mostly six to 10 pigs. The average period of gestation, or pregnancy, is 114 days. About an eight month old pig can be pregnant. PSY (pigs sow year), which is the number of pigs reproducible by a mother pig a year, is as many as 25 in well managed farms (Choi, J.S., 2002). One of the most severe and troublesome disease to which hogs are subject is hog cholera. Hogs are also troubled with a number of parasites such as the intestinal worm, the mange mite and with germ diseases such as influenza and erysipelas (Britannica Junior Encyclopedia, 1979).

As farms are getting bigger, a disease can kill all animals at a time, and preventive measures have been common. Antibiotics and growth promoters are mixed with feed resulting in over use (abuse) of medications, which consequently stimulate growth of antibiotic resistant organisms in the environment. Thus, there are some trends not to eat such a meat, and a numbers of farms not using antibiotics in feed are also growing. Preventive measures recommended are to raise pigs on pasture or fresh ground and keep their shelters clean and disinfected (Britannica Junior Encyclopedia, 1979).

Pig house are periodically cleaned with disinfectants and water about three times a week in Korean practices. The washing water is discharged with the feces and urine. The combined wastes have been normally used as nutrients for agriculture, but the amount produced actually exceeds the amount needed near the production sites, and requires transportation to a distant disposal site or on-site treatment based on a nutrient management plan.

In earlier practice (as shown in Table 1.1), nitrogen load was a guide for land application for agriculture, but later it was found that phosphorus was accumulated in soil and the excess phosphorus leached out of the soil to water bodies resulting in eutrophication. Furthermore, the phosphorus which accumulates in the soil reacts with iron to make a crystal so that the plant does not take up the iron as a nutrient. This is so called iron deficiency. For these reasons, phosphorus became a regulated load for land application. Thus, the amount of nutrient applicable will depend on soil characteristics and current nutrient levels. From this plan, it could be determined how piggery waste is to be managed, i.e., how much it can be applied to land.

It has not been a common practice to treat piggery waste and discharge the treated effluent into a receiving stream, since it is very difficult to achieve the effluent requirement. Rather, efforts have been made to reduce the amount of piggery waste which has been used as a chemical fertilizer. Needless to say, this is a sound approach. Even though the discharge into a stream is inevitable, the effluent must comply with the effluent regulations.

Table 1.1 Limits of nitrogen and phosphorus application to land.

Eco-Label (86/278/EEC)*		LFS MINAS** (2002)
170 kg N/ha (6.5)	→	100 kg N/ha (11.6)
60 kg P_2O_5/ha (1)	→	20 kg P_2O_5/ha (1)
120 kg K_2O/ha (3.8)		

* Official journal of the European Community (OJ, 1998) or Brinton (2000).
** Levy free surpluses within the mineral accounting system in Dutch (Ondersteijn *et al.*, 2002).
( ) denote N:P:K ratios

Table 1.2 gives the current regulatory effluent limits for piggery waste in Korea. As can be seen, the effluent limit for piggery waste is very comparable with municipal sewage and nightsoil (human excreta) treatment plants regardless the influent characteristics.

Table 1.2 Current regulatory effluent limits in Korea (mg/L) (Korean MOE, 2005).

Items	Farms for permit required: > 715 pigs		Farms for report required: 35 to 715 pigs		Centralized livestock wastewater treatment plants	Municipal sewage treatment plants	Nightsoil treatment plants
	Special areas	Others	Special areas	Others			
BOD	50	150	150	350	30	20	30
$COD_{(Mn)}$*	-	-			50	40	40
SS	50	150	150	350	30	20	30
TN	260	-	-	-	60	60	60
TP	50	-	-	-	8	8	8

*$COD_{(Mn)}$ values refluxed by $KMnO_4$. COD refluxed by $K_2Cr_2O_7$ is approximately 4 times $COD_{(Mn)}$

Piggery waste includes feces and urine. Feces include organic carbon (90% of the total discharge), and about 30% of the nitrogen and 80% of the phosphorus discharged. Plants need organic carbon, nitrogen and phosphorus. However, the organics in the feces require stabilization. Composting is a common process used for this purpose. During composting, organics are stabilized and aerobic microbes release heat to kill pathogens. After curing, the composted materials are applied to land. If a pig farm is too large, the composted materials will have to be transported to locations farther away from the producing site. This suggests a limit of farm size for sustainable composting or other means of feces management. One of the other means involves the use of lagoons which can store the whole waste including feces and urine, and give enough time to degrade naturally within a series of lagoons. The first lagoon becomes septic, degrading anaerobically. Other lagoons, following the first lagoon, approach aerobic treatment and accomplish nitrification. The effluent can be used for irrigation and the resulting sludges can be a source of nutrients for plants. During this treatment, odor problems from septic conditions and flooding problems from heavy storms are created. In the USA, embankment heights are increased to prevent the flooding. There are efforts to reduce the odor problem from lagoons, which include installation of covers or applying spray of nitrified liquid over the lagoon surface (Bull L S, 2005). Also, there are some efforts to prevent ground water pollution from the wastes leaked from lagoons (ASCE, 2003).

A Korean farmer built a multistory pig house with elevators to raise pigs in a small area. Areas with limited land can not utilize lagoon type treatment. However, the bioliquid or biofertilization process has been used. This technology is similar to aerobic digestion used in municipal wastewater treatment. It includes the aerobic degradation of organics at higher temperatures using the heat released from the aerobic degradation process like in aerobic composting. Unlike composting, this does not require a bulking agent. Piggery waste has strong organic materials which can be degraded to increase the reactor temperature. The aerobic digestion in a municipal plant is normally with secondary sludge (wastes activated sludge) which contains mostly nonbiodegradable fractions resulting in a lower reactor temperature. The higher reactor temperature is good to kill the pathogens and keep the nitrogen content in the liquid rather than lose it through nitrification and denitrification. This practice also requires a long transportation distance when the animal farm size is too large for local processing.

This book includes seven chapters. Chapter 2 describes piggery waste characteristics which depend on climate and feed. Chapter 3 describes solid separation and overall management systems for recycling technology (Chapter 4) and nutrient removal (Chapter 5). Biological nutrient removal (BNR) systems can remove organics while removing nutrients, but there are leftovers such as nonbiodegradable COD, which can be a target for removal when the effluent discharges to a nearby water source or when a drinking water supply is required for animals. Chapter 6 gives some illustrations on the removal of such residual pollutants by advanced oxidation, adsorption and membrane technology. Chapter 7 introduces integrated management systems. This book does not deal with odor problems from piggery waste management. However, readers can acquire a basic knowledge of piggery waste management for present and future needs.

REFERENCES

ASCE (2003) *Animal waste containment in lagoons* edited by Reddi L. N., ASCE manuals and reports on engineering practice No. 105

Banks C. J., Salter A. M., Chesshire M. (2007) Potential of anaerobic digestion for mitigation of greenhouse gas emissions and production of renewable energy from agriculture: Barriers and incentives to widespread adaptation in Europe, *Wat. Sci. Technol.* **55**(10), 165-173.

Brinton W. F. (2000) *Compost quality standards & guidelines*, Woods End Research Laboratory, Inc. prepared for New York State Association of Recycle.

Britannica Junior Encyclopedia (1979). Hog. 171-173, Encyclopedia Britannica, Inc., Chicago, London, Toronto.

Bull L. S. (2005) *Innovative sustainable systems utilizing economical solutions (ISSUES),* final report, North Carolina State University, Raleigh, NC.

Campagnolo E R, Johnson K R, Karpati A, Rubin C S, Kolpin D W, Meyer M T, Esteban J E, Currier R W, Smit K, Thu KM, McGeehin M (2002) Antimicrobial residues in animal waste and water resources proximal to large-scale swine and poultry feeding operations, *Sci Total Environ,* **299**, 89-95.

Choi, J.S. (2002*) Selection of right pigs and pig farm management*, Hyunchuk Pub Co., Korea.

Korean Ministry of Environment (2005) *Effluent regulations for wastewater treatment plants.*

Ondersteijn C. J. M., Beldman A. C. G., Daatselaar C. H. G., Giesen G. W. J. and Huirne R. B. M. (2002) The Dutch mineral accounting system and the European nitrate directive: Implications for N and P management and farm performance, *Agriculture, Ecosystems and Environment.* **92**, 283-290.

2

Piggery Waste Characteristics

2.1 LIFE CYCLE OF PIG

It is essential to know about the life of pigs (Choi, J S, 2002) for the development of a proper waste management plan. There are three different kinds of pig farms: Weaning only, farrow to finish and combined. Figure 2.1 shows the growth stages of a pig. A baby pig grows faster in the first two months after birth and somewhat slower afterwards. It then approaches a stationary phase where food is consumed without an increase in the pig's weight. The weight of the live pig prior to killing is about 100 to 110 kg and is called the finished live weight (FLW).

In the early stages a high caloric food with high protein is supplied to the pigs, but the amount of food, its caloric values and the contents of protein are controlled during the latter stages to control the meat quality (accordingly, the pig's waste strengths will vary). Figure 2.2 shows water demand and food consumption for various weights of pigs. The daily water demand is about 4 to 5 L/head and the daily food consumption is between 2 to 3 kg/head for adult pigs. Needless to say, pig farms operate to harvest the maximum amount of meat with

 ISBN: 9781843391319. Published by IWA Publishing, London, UK.

minimum feed. The feed requirement also varies with room temperatures. Adequate temperatures are near 30ºC for baby pigs and 20ºC for adults. If the temperature is cooler, more food is required with less water demand, resulting in a stronger waste discharge. Conversely, higher temperatures produce a weaker waste.

Figure 2.3 shows how much each part of a pig contributes to a 100 kg finished live weight. From 100 kg of FLW, harvested bacon and meat are 21 kg and 34 kg, respectively. Chemical compositions are 3 kg ash, 27.7 kg fat, 13.9 kg protein and 50.4 kg water in 95 kg of total weight of 100 kg, which includes 5 kg of undigested food remaining when it is killed.

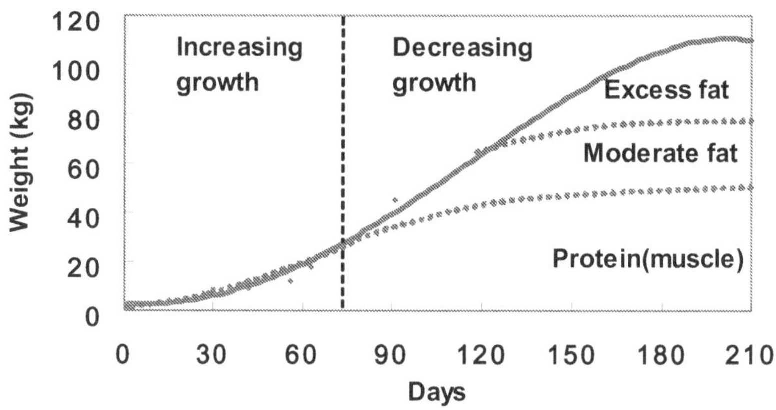

Figure 2.1 Body weight growth of pig with time.

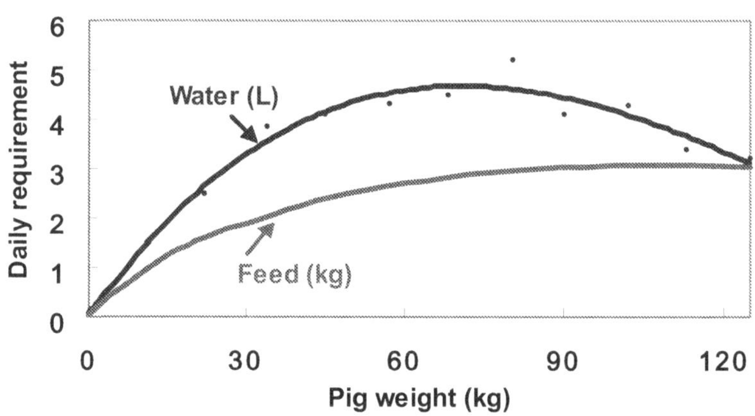

Figure 2.2 Daily water and food required for various weights of pigs (L and kg/head/d).

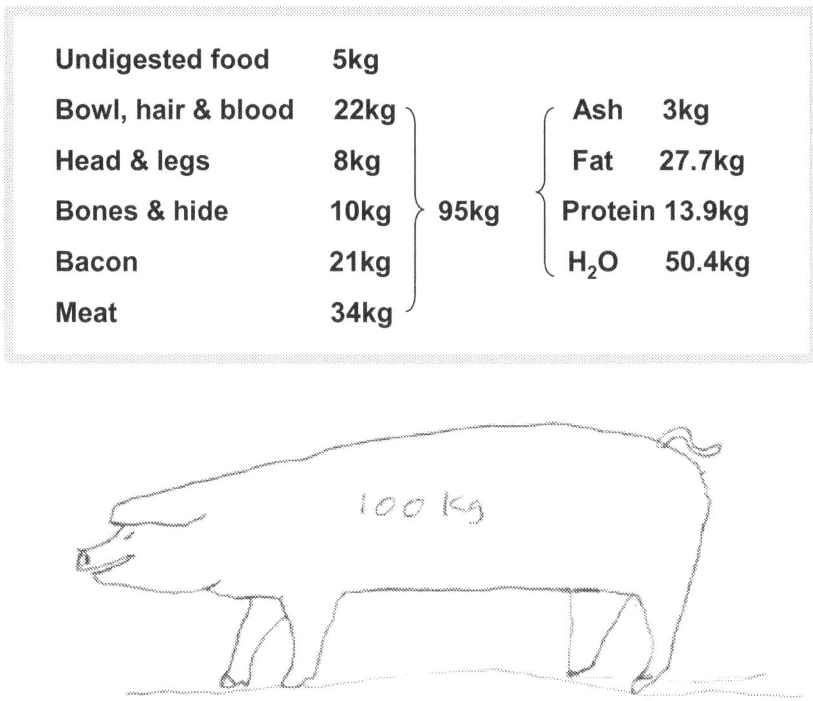

Figure 2.3 Composition of finished live weight (modified from Choi, 2002).

2.2 FOOD FOR PIGS (DIETS FOR PIGS)

Some of the nutrients provided for animals will be discharged to the environment as excreta and urine which creates an environmental problem. When the farm size was small, the waste problem was not recognized since the waste could be applied to an agricultural area where the pollutants were degraded and used by soil bacteria and plants. Actually, food waste from human was used for animal feed. As the farm sizes increased, the waste problem increased and received more recognition. Today human food waste is no longer used as a major source of feed for animal. This is particularly true for larger animal farms and where a high quality of meat is required. The larger animal farms are feeding animal foods produced form food suppliers.

When the size of the farm was small, different kinds of animal food including human food wastes were fed to the animals and the composition of the animal wastes varied depending on type of food fed. However, the industrialized animal farms require a uniform food composition to produce a higher quality of meat product and resulting in a waste with similar characteristics. There are some diet differences between ages and types of pigs, since the amount and types of feed vary with animal ages or weight of animal (pig or swine in this case) as shown in Table 2.1 (National Research Council, 1998).

It can be seen that more food energy is required as the animal grows. The food energy is 3,400 kcal/kg feed and about 7,200 kcal/day is required for a 50 kg pig resulting in 2.1 kg of feed. Minerals and vitamins are added to nutrients, which include iron, zinc, calcium, phosphorus and copper as macronutrient demand. Antibiotics or promoters can also be fed with the food to prevent disease (This will be discussed in Chapter 4).

Table 2.2 represents the characteristics of a food prepared for different growing stages of pigs at a Korean farm. The feed was analyzed to determine parameters commonly used in the environmental engineering field. For instance, COD (chemical oxygen demand), TS (total solids), VS (volatile solids) and other parameters were measured. TS and VS contents/kg feed are more consistent regardless of age, but COD, TN and TP contents/kg feed vary. Higher COD foods are fed to pregnant and mother pigs, higher nitrogen foods are fed to newly borne pigs, and higher phosphorus foods are fed to newly borne and mother pigs. COD/VS ratios are 0.68 to 1.8 with an average of 1.17, and COD/TN ratios are 4.6 to 29.8 with an average of 15.8. TN/TP ratios are 3.5 to 15.8 with an average of 6.2. Chloride requirements do not vary greatly for different ages of pigs.

Table 2.2 Characteristics of cereals for Korean pigs at different ages (g/kg).

Purposes	TS	VS	COD	TN	TP	Cl^-
Nursery #1	890	830	954	73	13.0	3.7
Nursery #2	890	673	852	38	9.4	3.39
Weaning	885	847	584	127	80	3.75
Pregnant	890	839	1,150	52	9.7	3.72
Sows	860	804	1,458	49	14.4	3.50
Finishing	880	837	647	29	7.3	3.49
Farrow	885	838	993	49	5.7	3.5
Average	882	810	948	60	9.7	3.58

Table 2.1 Daily nutrient requirements of growing pigs (90% dry matter)[a].(Modified from NRC, 1998).

	Body weight (kg)					
	3-5	5-10	10-20	20-50	50-80	80-120
Average weight in range (kg)	4	7.5	15	35	65	100
DE content of diet (kcal/kg) [b]	3,400	3,000	3,400	3,400	3,400	3,400
Extended DE intake (kcal/day)	855	1,690	3,400	6,305	8,760	10,450
Extended food intake (g/day)	250	500	1,000	1,855	2,575	3,075
	Mineral requirements (amount/day)					
Calcium (g)[c]	2.25	4.00	7.00	11.13	12.88	13.84
Phosphorus. total (g)[c]	1.75	3.25	6.00	9.28	11.59	12.30
Phosphorus. available (g)[c]	1.38	2.00	3.20	4.27	4.89	4.61
Sodium (g)	0.63	1.00	1.50	1.86	2.58	3.08
Chlorine (g)	0.63	1.00	1.50	1.48	2.06	2.46
Magnesium (g)	0.10	0.20	0.40	0.74	1.03	1.23
Potassium (g)	0.75	1.40	2.60	4.27	4.89	5.23
Copper (mg)	1.50	3.00	5.00	7.42	9.01	9.23
Iodine (mg)	0.04	0.07	0.14	0.26	0.36	0.43
Iron (mg)	25.00	50.00	80.00	111.30	129.75	123.00
Manganese (mg)	1.00	2.00	3.00	3.71	5.15	6.15
Selenium (mg)	0.08	0.15	0.25	0.28	0.39	0.46
Zinc (mg)	25.00	50.00	80.00	111.30	129.75	153.75

Note: a Pigs of mixed gender (1:1 ratio of barrows to gilts). The daily requirements of certain minerals may be slightly higher for pigs having high lean growth rates (325 g/day of carcass fat-free lean), but no distinction is made. Vitamin requirements are omitted from the original table.

b DE (digestible energy including the energy discharged with urine and gaseous products of digestion).

c The daily amounts of calcium, phosphorus, and available phosphorus are slightly higher in developing boars and gilts from 50 to 120 kg body weight.

2.3 WASTE DISCHARGE FROM PIG FARMS (PIG HOUSES)

Figure 2.4 shows the amount and strength of waste discharged from different weights of pigs (Korean RDA, 2002). The maximum waste flow occurred with about 50 kg pigs, but the maximum pollution loads of BOD, nitrogen and

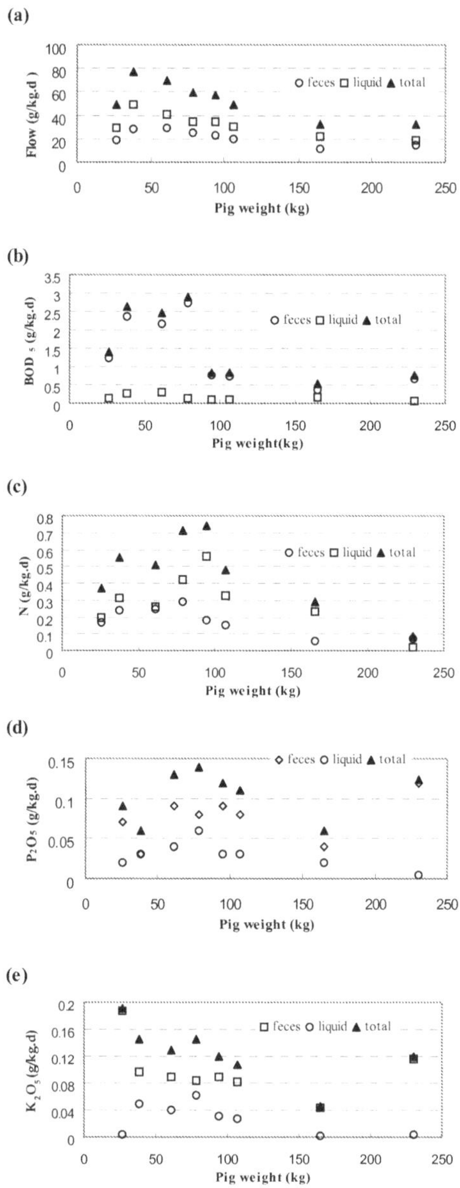

Figure 2.4 Waste discharge and pollution loads from pigs based on weights (modified from Korean RDA, 2002): (a) Flow (b) BOD (c) Nitrogen (d) Phosphorus (as P_2O_5) (e) Potassium (as K_2O).

phosphorus occurred from pigs with about 75 kg. Thus the waste volume and its strength vary with the weights of pigs, how to grow the pigs and how to manage the waste stream. Baby pigs will grow to be adult pigs weighing about 100 kg within 6 months. The adult pigs will be sold for meat harvesting.

The feces and urine can be collected together or separately depending on how the waste is to be used or treated. If feces are collected separately for composting, the liquid (mostly urine) will be the wastewater to be treated. If the feces are not separated, then the whole waste can be considered as wastewater. Depending on labor costs, the separation methods will be by either manual or mechanical means. In general, small farms manually separate feces with shovels, resulting in only a small amount of feces being discharged with urine. Big farms use mechanical separators such as scrapers to remove feces. Farms in the USA, in most cases, do not separate feces, because they intend to recycle the whole waste following treatment with a series of lagoons.

Separation of feces allows weaker wastewater discharge, but the collected feces must be used for specific purpose such as application to agricultural land after storage or composting. Some Korean farms do not separate feces from urine, because they are utilizing the whole waste as liquid fertilizer (bioliquid) for agricultural areas. Therefore, whether the separation of feces is employed or not will be determined by the final disposal or utilization method.

Previously, animal manure and human excreta were used for agriculture and were considered as "yellow gold," but today this attitude, in most cases, has changed because there are limits on waste application. The small animal farms are getting bigger and industrialized with limited land, resulting in serious ground and surface water pollution problems resulting from increased animal wastes from the confined animal feeding operations (CAFOs). In most cases, agricultural areas adjacent to CAFOs are saturated with nutrients. Consequently, Korea is facing increasing animal waste disposal problems. Fortunately, demands for organic fertilizers have been increasing; hence, the opportunity of recycling pig waste is gradually increasing.

Table 2.3 presents how much manure and nutrients are generally produced from beef and dairy cows, pigs and chickens. These values may not be representative, since they vary among farms. However, they do show that pig waste contains more nutrients than beef and dairy manures. Pigs generally discharge 2 L of feces and 3 L of urine. The amount of urine discharge is related to climate since pigs drink more water in hot climates, as previously described.

Table 2.3 Manure discharge and its fertilizer content based on adult animal (Korean National Livestock Research Institute, 2000)

Animals	Weight (kg)	Maure (kg/head/day)					Fertilizer content (%)								
		Feces		Urine		Total	N			P_2O_5			K_2O		
		Wet kg	% water	Wet kg	% water	Wet kg	Feces	Urine	Total	Feces	Urine	Total	Feces	Urine	Total
Beef	350	11.4	81	3.9	95.4	15.3	0.34	0.45	0.28	0.25	0.006	0.08	0.09	0.47	0.19
Dairy	450	24.3	83.9	10.2	95.1	34.5	0.26	0.34	0.37	0.11	0.003	0.19	0.14	0.31	0.19
Pig	60	1.8	74.7	2.9	98.4	4.7	0.85	1.02	0.95	0.37	0.07	0.25	0.23	0.28	0.26
Chicken				-	-	0.11		-	0.01		-	0.37		-	0.46

Note: $P_2/P_2O_5 = (31\times2) / (31\times2+16\times5) = 0.44$, $K_2O = K_2/K_2O = (40\times2) / (40\times2+16) = 0.83$

Table 2.4 Manure production and composition from different animals.

	Daily production				Composition				
	Per 1,000 lb Live animal		Per 500 kg Live animal		BOD	Volatile solids	N	P_2O_5	K_2O
Adopted from NAS (2001)									
	Wet		Wet		Percent of wet weight				
Animal	Volume (ft^3)	Weight (lb)	Volume (ft^3)	Weight (lb)					
Dairy cattle	1.33	76.9	0.038	38.5			7.98	0.38	0.10
Beef cattle	1.33	83.3	0.038	41.7			9.33	0.70	0.20
Pig	**1.00**	**56.7**	**0.028**	**28.4**			**7.02**	**0.83**	**0.47**
Sheep	0.70	40.0	0.020	20.0			21.5	1.00	0.30
Poultry	1.00	62.5	0.028	31.3			16.8	1.20	1.20
Horses	0.90	56.0	0.025	28.0			14.3	0.86	0.13
Adopted from Korean RDA (2002)									
Dairy catttle	1.27	88.2	0.036	40.0	1.40	5.00	0.27	0.07	0.18
Beef cattle	0.53	47.2	0.015	21.4	2.00	7.23	0.37	0.20	0.20
Pig	**1.42**	**88.9**	**0.040***	**40.0**	**0.75**	**5.20**	**0.75**	**0.17**	**0.22**

*Excluding cleaning water which is estimated to be about 10 L/d or 1 L/head/d.

Note: 1 animal unit (AU) = 500 kg which is considered to be about 10 pigs from farrow to finish.

Table 2.4 presents manure production and its composition for different animals based on 500 kg, which is defined as an animal unit (1 AU). The NAS (2001) data originated from a 1971 survey, while the Korean RDA (2002) data are fairly recent. Note that the waste discharge ranges from 4 to 8% of the animal weight. Table 2.5 compares the chemical composition of urine for human and animals based on 100 kg N.

Table 2.5 Difference in chemical composition between human and animal urine based on 100 kg N.

Constituents (kg)	Human[(1)]	Cattle[(1)]	Pig urine[(2)]
Tot-N	100	100	100
NH_4-N	83	52	55
K^+	29	294	22
Na^+	40	6.1	35
Ca^{2+}	<0.05	0.09	5
Mg^{2+}	<0.015	1.92	2
P	6.2	0.77	8.3
Cl^-	57	5.8	35
SO_4^{2-}	2.4	2.3	5.6
HCO_3^-	165	364	172

[(1)] Imake F and Hermann H. H (1998), [(2)] Collected values for this book

It is estimated that urine constitutes about 65% of the N and 23% of the P in the total discharge (Lee, 2006). One pig generates daily about 2 L of feces and 3 L of urine with 510 g COD, 30 g N and 4.8 g P based on an average pig weight of 50 kg/head as shown in Figure 2.5 and Table 2.6. The combined piggery waste concentration including feces and urine can be estimated to be 102 g COD/L, 6 g N/L and 0.96 g P/L. When feces are separated for agricultural reuse, the remaining wastewater (still containing some of remained feces) exhibits 18 to 41 g COD/L, 2 to 3.6 g N/L and 0.3 to 0.5 g P/L, 0.26 to 0.48 g Ca^{2+}/L, 0.11 to 0.16 g Mg^{2+}/L which would be the representative design concentrations as shown in Figure 2.5, 2.6 and 2.7.

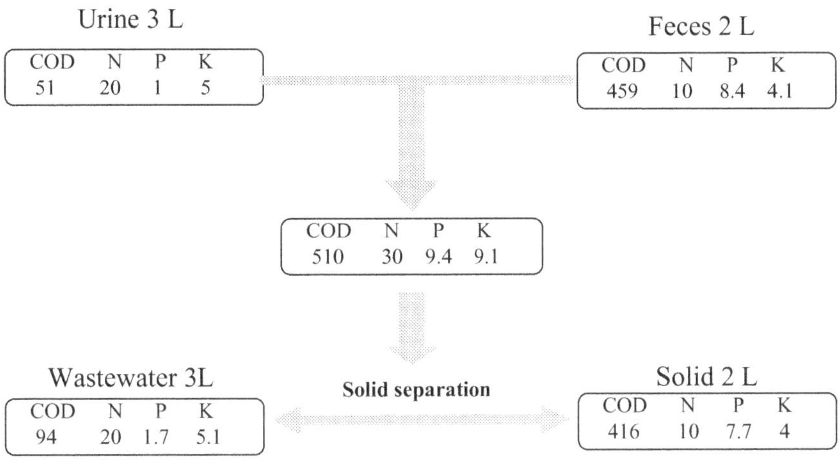

Figure 2.5 Proposed wastewater loads for design in g/head/d after proper solid separation.

Table 2.6 Proposed daily discharge per head per day for design in comparison with human waste.

Items	Pigs			Human*		
	Feces	Urine	Total	Feces	Urine	Total
Flow(L/d)	2	3	5	0.25	1.25	1.5
COD (g/d)	459	51	510	40	10	50
(%)	90	10	100	80	20	100
BOD (g/d)	230	25	255	20	5	25
(%)	90	10	100	80	20	100
N (g/d)	10	20	30	1.4	12	13.4
(%)	34	66	100	10	90	100
P (g/d)	7.2	2.2	9.4	0.8	1	1.8
(%)	77	23	100	45	55	100
K (g/d)	4.1	5	9.1	0.6	2.6	3.2
(%)	45	55	100	19	81	100

* Choi E (2000)

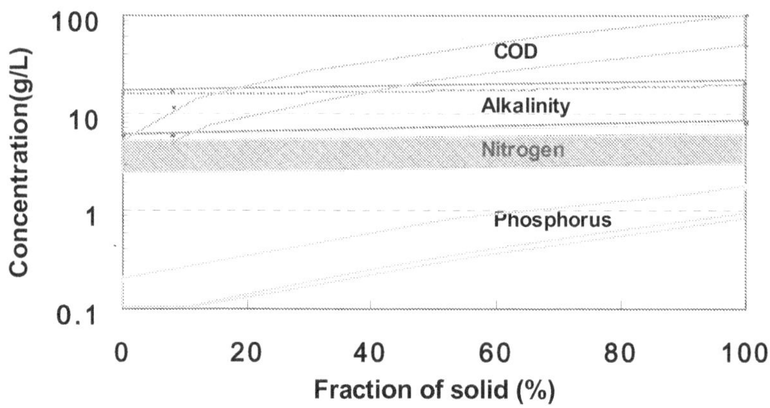

Figure 2.6 Estimated piggery wastewater concentrations with different solid contents and climate conditions all year around. Higher concentrations are from winter operation.

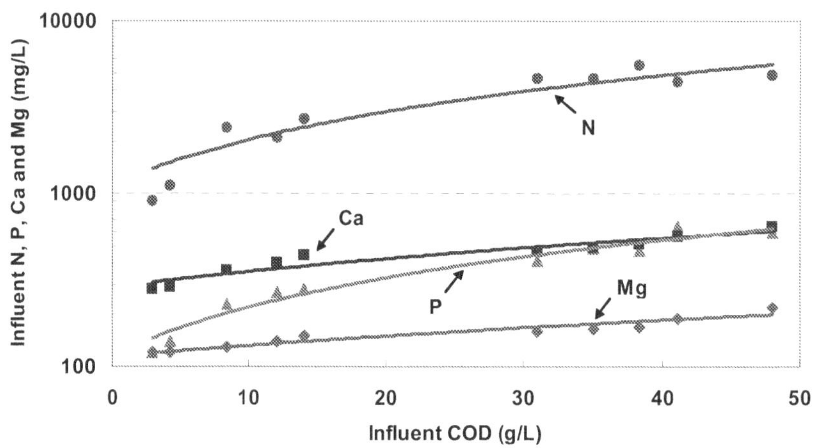

Figure 2.7 Relation of influent COD and N, P, Ca and Mg contents in piggery waste.

A recent study at North Carolina State University (Bull, 2005) indicates that the calculated average nutrient excretion, based on inflow and outflow data for different houses containing 6,480 animals, are different from the data summarized by ASAE (2003). These data are given in Table 2.7.

Table 2.7 Average nutrient excretion per pig (g/pig/day) from NC State University and ASAE.

Sources	Total N	TKN	NH_4-N	NO_3-N	P	Zn	Cu	Solids
Bull (2005)	25	25	15.4	<1	7.2	0.4	0.06	200
ASAE (2003)	39					6		466

In addition, the nitrogen fraction can be computed from the chemical composition of manure ($C_6H_{13}NO_5$) given by Hill and Barth (1975). This gives a nitrogen content of approximately 9% of the VS with a COD/VS ratio of about one. If the nitrogen content of manure is 30 g/pig/d, the COD discharge, according to the following Equation, will be 342 g/pig/day, which is far smaller than the proposed design value of 510 g/head/d.

$$C_6H_{13}NO_5 + 6O_2 \rightarrow 6CO_2 + 5H_2O + NH_3 \quad (2.1)$$

In addition, the wastewater is basically urine itself. Water demands by pigs vary with climate; temperature and humidity. When weather is hot, pigs need more water. Conversely, water demand is reduced during winter. Also, there is a water demand for cleaning the pig pens. The summer demand increases to about 1.8 times the winter demand in Korea. In short, temperature effects result in added flow and concentration fluctuations originating from the differences in the pig's weight and diet. Figure 2.6 shows variations of wastewater concentrations depending on solid separation from sources. The total COD of combined feces and urine is about 100 g/L and the wastewater concentrations decrease as solid are separated from the combined streams. Alkalinity and nitrogen concentrations do not vary greatly, but phosphorus concentrations vary greatly with solid contents. Table 2.8 shows measured influent characteristics of full scale plants in Korea.

Table 2.8 Measured influent characteristics of pilot and full scale treatment plants in Korea* (mg/L).

Constituents	Pilot Plant (1)	Full Scale (2)	Full Scale (3)
TSS	14,289		
VSS	10,640		
TCOD	31,230	45,320	41,300
SCOD	20,482		28,300
TKN	4,017	6,680	5,500
NH_4N	2,515	3,470	3,400
TP	515	670	
SP	362		
PO_4P	284		
Alkalinity	11,092	8,560	19,000
pH	8.4	8.3	8
COD/N	7.8	6.6	7.5
ICOD/VSS	1.2		
Mg^{2+}	124		
Ca^{2+}	286		
K^+	762		
SO_4^{2-}	336		

*Includes some solids even though solids separation was practiced. Full scale units are centralized facilities. The concentrations in full scale units include the contribution from the reject water discharged from sludge treatment.

Note: TCOD = total COD = soluble COD (= SCOD) + insoluble or particulate COD (= ICOD), SP = soluble phosphorus

REFERENCES

ASAE (2003) *Proceedings of the 9th International Animal, Agricultural and Food Processing Wastes*, Oct 12, Research Triangle Park, NC, 257-262.

Bull L S (2005) *ISSUES*, Final report, NC State University

Choi E (2000) *Water supply and pollution control*, Cheongmoongak Publishing Co. Seoul, Korea (in Korean).

Choi, J.S. (2002) *Selection of right pigs and pig farm management*, Hyunchuk Pub Co., Korea.

Hill D T and Barth C L (1975) Simulation of fundamental anaerobic lagoon kinetics. In: *Processing and Managing Agricultural Waste. Cornell Agricultural Waste Management Conference Proceedings.*

Imake F and Hermann H. H (1998) Characterization of the municipal wastewater part human urine and a preliminary comparison with liquid cattle excretion, *Water. Science. Technology.* **38**(6), 9-16.

Korean National Livestock Research Institute (2000) *Research report on livestock*, #32203, Korean RDA.

Korean Rural Development Administration (2002) *Guidelines for the use of liquid fertilizer.* Technology Transfer 11-1390093-000082-01, www.rda.go.kr

Lee J H (2006) *Piggery waste treatment with ABR and BNR processes*, MS thesis, Department of Civil and Environmental Engineering, Korea University, Seoul, Korea

National Academy of Sciences (2001) *Methane generation from human, animal, and agricultural wastes*, Reprinted from 1977 edition by Books for Business, New York, Hong Kong.

National Research Council (1998) *Nutrient requirements of swine*, tenth revised edition, Subcommittee on Animal Nutrition, Board on Agriculture, The National Academy of Sciences.

3

Solids Separation and Overall Management Systems

Solids separation is related to an overall management system as shown in Figure 3.1. Solids separation reduces waste concentrations in the liquid which would require larger tank volumes for treatment if the solids were not removed. The concentrated waste from solids separation increases the potential of recycling with composting. If anaerobic digestion is to be utilized, there is no reason to separate the solids as European and US experiences have shown (discussed later in Chapters 3 and 7). The major recycling technology at present is energy recovery with anaerobic digestion and land application of treated sludge. As shown in Table 1.2; however, the applicable rate was previously based on nitrogen, but phosphorus levels created problems in the soil and streams when leached from the soil. Therefore, the applicable amounts of solids have been reduced.

by Euiso Choi. ISBN: 9781843391319. Published by IWA Publishing, London, UK.

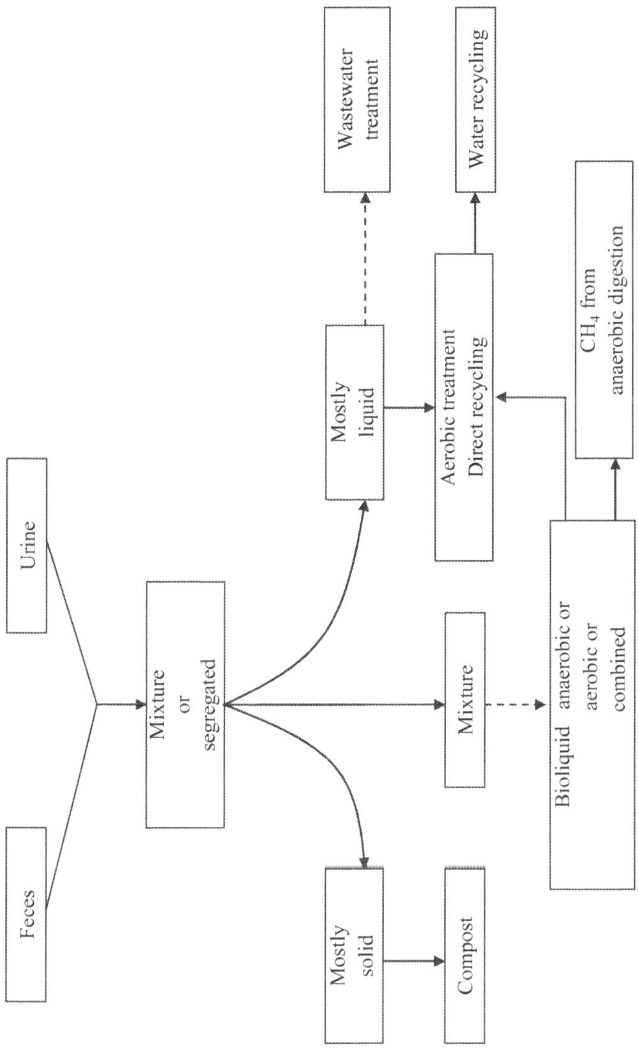

Figure 3.1 Overall piggery waste management system.

3.1 TREATMENT OF ANIMAL WASTE

Confined animal feeding operations typically require the farmer or landowner to either acquire more land to accept waste materials for land application or to initiate some forms of waste treatment to minimize the volume of waste. Complicated and costly methods of waste management system are generally not feasible on the farm due to labor and economic constraints. However, waste management will be more attractive if it can produce some valuable materials to offset the cost required for treatment.

3.2 SETTLING AND DIGESTION

Gravity settling basins or unmixed anaerobic digesters can be used to separate solids/liquids through the natural settling process (Haelzle, 2005). The structure is designed to allow solids which settle to be digested. It was found that thickening and digestion could not be separated. A study by North Carolina State University (Bull, 2005) to install a thickener prior to feeding to an anaerobic digester failed. It was found that gas was produced in the thickener which caused solid separation to fail. Both settling and digestion can occur simultaneously.

3.3 MECHANICAL SEPARATORS

There are many different kinds of mechanical separators commercially available. The simplest separator involves the screening of the waste stream with different sizes of openings. Some means of scrapers are used to remove the accumulated solids. Other more complicated mechanical devices are also used. Dissolved air flotation with chemical addition as well as centrifugation is also used.

3.4 ANAEROBIC BAFFLED REACTOR (ABR)

Figure 3.2 shows a schematic diagram of a combined system of settling and anaerobic digestion named ABR (anaerobic baffled reactor). It was tested under laboratory and field conditions to determine its efficiencies with different retention times and temperatures. Laboratory units were operated at constant temperatures with and without covers to determine how much COD was removed by anaerobic digestion. This system was developed by Korea University in Seoul, Korea to regulate feed concentrations for an aerobic biological process during summer and winter periods.

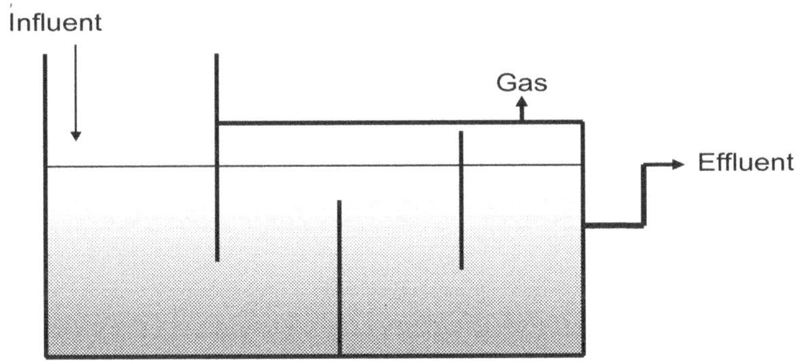

Figure 3.2 Conceptual configuration of anaerobic baffled reactor with cover.

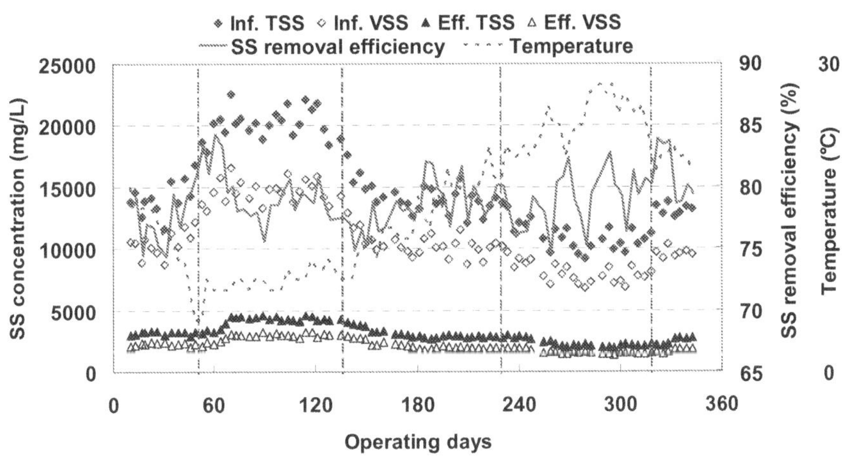

Figure 3.3 TSS and VSS removal efficiencies with temperature for an ABR pilot plant system (Yu, 2007).

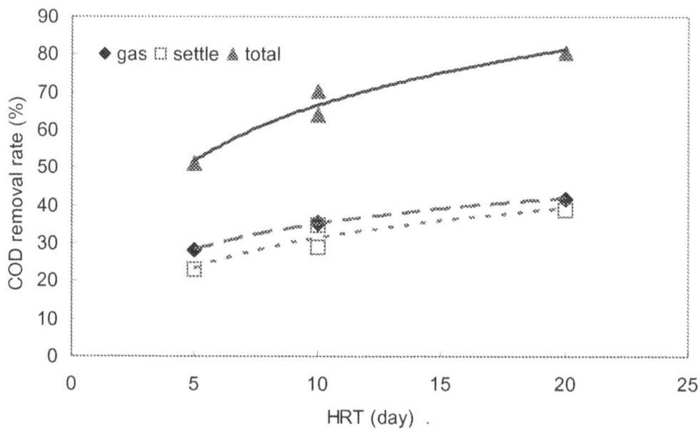

(a) COD

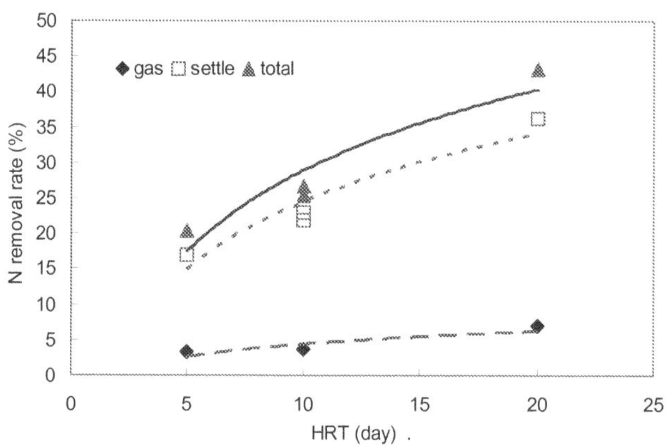

(b) Nitrogen

Figure 3.4. COD and nitrogen removal from a laboratory ABR operation at 35°C (Lee, 2006).

Figure 3.3 shows total suspended solid (TSS) and volatile suspended solid (VSS) removal efficiencies with temperature variations of 7 to 28°C. The removal efficiencies were 75 to 85% regardless of the temperature range. Figure 3.4 shows how COD and nitrogen were removed in an ABR system operated at 35°C. COD was removed both by settling and anaerobic digestion. Anaerobic digestion resulted in a methane production of 0.32 m^3 CH_4/kg COD removed. At a 10 day HRT (hydraulic retention time), the COD removal efficiency was about 70%, of which 35% was by settling and the other 35% was by anaerobic digestion. Nitrogen removal in an ABR system was considered by volatilization based on a mass balance of nitrogen (Lee, 2006).

Table 3.1 summarizes COD, N and P removal efficiencies of an ABR system at different temperatures. As indicated previously, the waste concentrations vary with temperature which affects the amount of wastewater produced. The higher ambient temperature produces more wastewater to dilute the waste concentration. Thus, the HRT will be shortened at higher temperature. Figure 3.5 shows examples how to design for removal efficiencies at different temperature and HRTs. COD removal efficiencies vary from 30% at 15°C with 10 day HRT to 50% at 35°C with 5.5 day HRT. Nitrogen removal efficiencies vary 10 to 20% for the same conditions. Phosphorus removal is about 55% .regardless of temperature and HRT.

Table 3.1 COD, N and P removal efficiencies at different temperatures of an ABR.

Temperature (°C)	Removal efficiencies (%)		
	COD	N	P
15	35	10	55
25	45	15	55
35	64	22	54

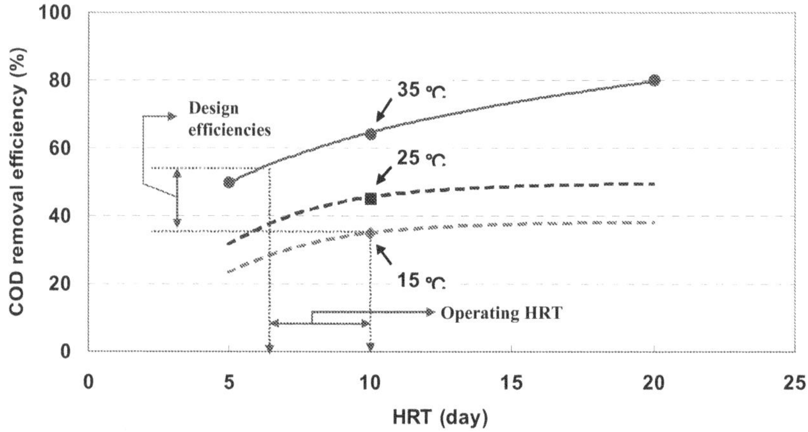

(a) COD

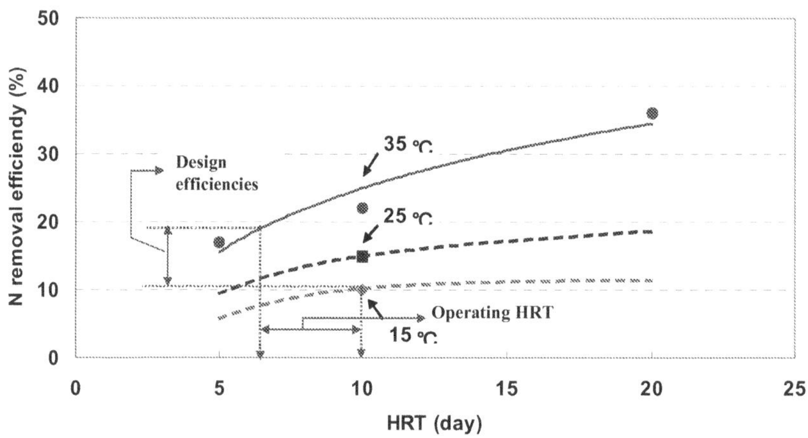

.(b) Nitrogen

Figure 3.5 Determination of design efficiencies of COD and nitrogen removal at actual condition (ABR).

Figure 3.6 shows fractionated COD and nitrogen using an IWA format before and after ABR operation at different temperatures. The raw COD and nitrogen concentrations are 18,000 to 32,000 mg/L and 2,000 to 3,600 mg/L, respectively, after removal of feces with mechanical scrapers. After ABR treatment, COD concentrations are reduced to 10,000 mg/L during summer and 20,000 mg/L during winter, while nitrogen concentrations are reduced to 1,700 mg/L during summer and 3,200 mg/L during winter (Yu, 2007). The lower COD concentration during the summer improves the following aerobic biological process while not greatly increasing the reactor temperature. The volume of settled sludge to be removed from an ABR system will be about 7 to 24% of the inflow volume depending on temperature (Lee, 2006). Less sludge removal is required during the summer.

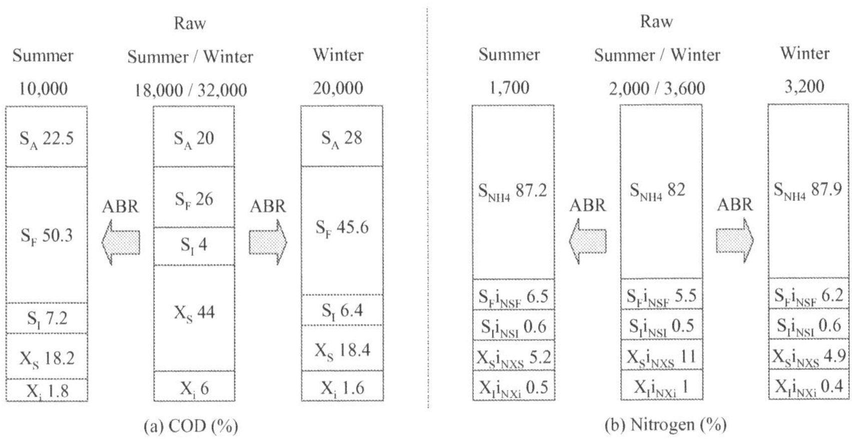

Figure 3.6 Fractionated COD and nitrogen before and after ABR operation. Raw concentrations designate the liquid stream concentrations after removal of feces with mechanical scrapers (Yu, 2007).

S_A = Fermentation products (acetate)
S_F = Fermentable readily biodegradable substrate
S_I = Inert, nonbiodegradable organics
X_S = Slowly biodegradable substances (particulate components)
X_i = Inert, nonbiodegradable organics (particulate components)
S_{NH4} = N content of soluble substrate
S_Fi_{NSI} = N content of inert solubles
X_Si_{NXS} = N content of particulate substrate
X_ii_{NXi} = N content of inert particulates

REFERENCES

Bull, L. S. (2005) Innovative sustainable systems utilizing economic solutions (issues), North Carolina State University, Raleigh, NC.

Haelzle, J.J. (2005) Concentrated animal feeding operations (CAFOs) and manure management – An alternative approach, Animal agriculture and processing: Managing environmental impact, Air & Waste Management Association, WEF, St. Louis, MO.

Lee, J. H. (2006) Piggery waste treatment with ABR and BNR processes, MS thesis, Department of Civil and Environmental Engineering, Korea University, Seoul, Korea.

Yu, Y. B. (2007) The dynamic nature of MLE process for piggery waste treatment with temperature, MS thesis, Department of Civil and Environmental Engineering, Korea University, Seoul, Korea.

4

Recycling Technology

4.1 INTRODUCTION

Pigs are among the first animals tamed by humans. Asiatic swine were tamed probably as early as 2900 BC and European varieties in 1500 BC (Britannica Junior Encyclopedia, 1979). Since that time, confined pigs are raised to harvest meat protein while the manure is used as a fertilizer for agriculture. Figure 4.1 presents a conceptual scheme of piggery waste management for sustainable pig farming. During the pig farming process, there are number of environmental problems; in particular, surface and ground water pollution as well as odor problems. If land area is large enough to receive the waste and grow plants, there will be minimal pollution problems. However, the sizes of farms or so called confined (concentrated) animal feeding operations (CAFO) are getting bigger; hence, the problem is increasing. The basic problem is that there is not enough land to accept the increased manure production.

The world grain harvest area has been reduced from 0.23 to 0.10 ha/person during the last 50 years (Brown *et al.,* 2000), while meat demand has increased about two times during the last 30 years from 9 to 19 kg/capita in 1961 to 1991. It will continue to increase to 30 kg/capita by 2025 as shown in Table 4.1 (Olutayo, 2005). Waste application to land has resulted in the over-accumulation of nutrients in the soil resulting in their discharge to water bodies causing eutrophication of lake waters and nitrate pollution problem in ground water.

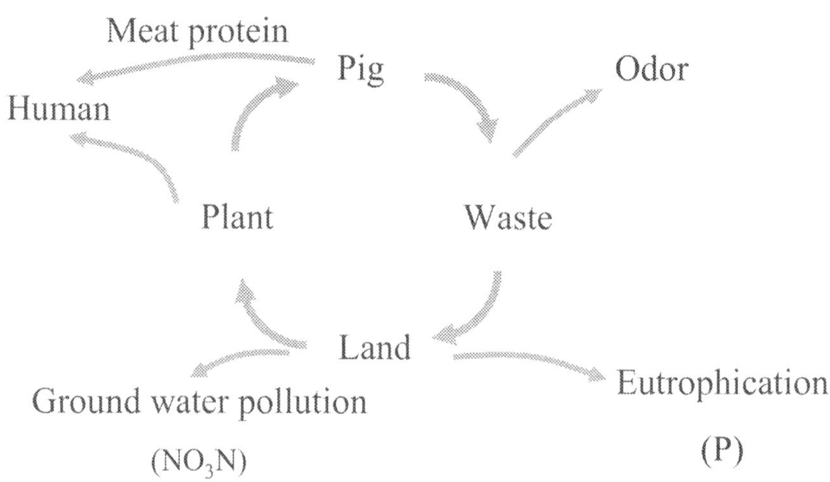

Figure 4.1 Animal waste and agriculture with related problems.

Table 4.1 Historical statistical 3-year average data and projections for the years 2010 and 2025 of food demand per capita, total domestic demand, self sufficiency ratio (SSR)and production for total meat and cereals for all developing countries including China. For cereals total per capita demand (including feed and other uses), the use as animal feed, and average yield are also presented (Olutayo, 2005).

Periods	1961/63	1969/71	1979/81	1989/91	2010	2025
Population						
10^6 Inhabitants	2,116	2,583	3,228	3,998	5,619	6,802
Meat						
Demand per capita (kg/cap.)[a]	9	11	13	17	25	30
Total demand (10^6 ton)	19	28	43	69	143	204
SSR	105	105	101	101	100	100
Production (10^6 ton)	20	29	44	70	143	203
Cereals						
Cereals total (kg/cap.)	171	191	219	237	257	278
Cereals food (kg/cap.)[a]	133	146	161	168	174	177
Domestic demand (10^6 ton)	361	494	706	940	1,443	1,844
SSR	97	97	92	92	90	88
Production (10^6 ton)	351	480	648	862	1,296	1,655
Use as animal feed (10^6 ton)	32	56	107	154	327	490
Use as animal feed (%)[b]	9	11	15	16	23	26
Feed intensity (kg/kg)[c]	1.5	1.8	2.1	1.8	1.9	1.9
Yield (ton/ha)	1.0	1.2	1.6	2.0	2.8	3.4

All projections 1990-2010 are updated from the FAO study "World Agriculture: Towards 2010" (Alexandratos N, 1995) using the UN (1994) scenario of population. Self-sufficiency ratios were assumed constant in 2010-2025, except for Near East and North Africa, where self-sufficiency ratios for cereals were lowered to prevent the arable and irrigated land areas from exceeding the potential areas.

[a] Projections of food consumption are based on estimated income elasticity of demand from the FAO study "World Agriculture: Toward 2010" (Alexandratos, 1995).

[b] Feed use as % of domestic demand.

[c] kg cereals/kg feed intensity weighted livestock production; the latter is computed as 0.3 (beef + mutton) + 0.1 (milk) + 1.0 (pork + poultry meat + eggs). This weighing is required since data on feed use for individual animal categories is not available from statistics.

4.2 STABILIZATION TECHNOLOGIES APPLIED TO PIGGERY WASTE FOR ANIMAL AGRICULTURE

Figure 4.2 shows a general scheme for choosing stabilization technologies for animal agriculture based on nutrient requirements of common crops. There are three pathways for reuse: composting, bioliquid and liquid treatment for discharge or reuse.

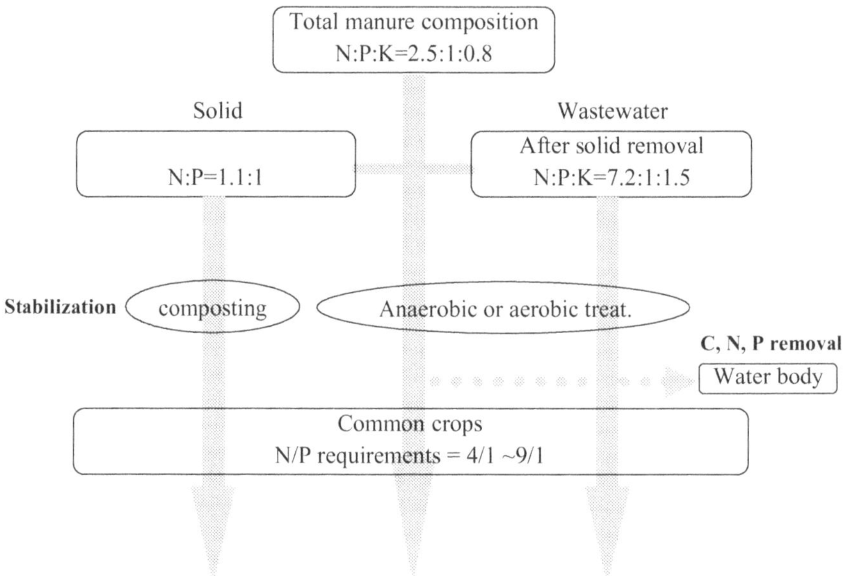

Possible environmental adverse impact	Excess P → Eutrophication	Excess P → Eutrophication	Excess N → Ground water problem (NO_3-N)
Practices desired	Add right bulking agent	Excess P removal by struvite	Excess N removal by stripping, nitrification and denitrification
To make better fertilizer	Add N	Add N	Add chemical fertilizer

Figure 4.2 Piggery waste: nutrient contents and measures to minimize adverse impacts on environment based on nutrient requirement of common crops.

Pig feces and urine are discharged as a mixture. Depending on how the waste is reused, the solid can be segregated manually or with scrapers and composted. The mixture without segregation can be used as bioliquid (liquid fertilizer) after digestion (either anaerobic or aerobic). Many plants in Europe use anaerobic digestion and recover the methane gas while the digested sludge is used for agriculture. Liquid treatment is used for those regions which do not have enough agricultural land. The factors to be considered for land application are as follows:

(1) Nutrient Values (N:P ratios)

The major plant nutrient requirements are C, N, P and K. Figure 4.2 presents fresh pig manure composition as N:P:K of 2.5:1:0.8 using a Korean pig feeding operation. Common crops require a ratio of N:P = 4:1 to 9:1 (Baker *et al.*, 1994). As shown on Table 1.2, European Eco-label standards specify applicable rates per ha of not more than 17 kg N, 6 kg P_2O_5 and 12 kg K_2O which are equivalent to a ratio of N:P:K of 6.5:1:3.8 (EC directive 86/278/EEC). Thus, it can be concluded that either nitrogen or phosphorus needs to be added or removed to make a proper N:P ratio. The fresh manure can not be directly applied to the land, since a large application of biodegradable carbon will consume dissolved oxygen from the soil, resulting in serious plant damage. Therefore, the easily biodegradable portion must be removed or stabilized prior to application. The common technologies used for this purpose are aerobic and anaerobic stabilization.

(2) Disinfectant

Various disinfectants are regularly applied to pig houses at a rate of 0.7 L/m^2 of floor. Generally, the regular dose does not affect subsequent treatment efficiencies; however, Hwang and Min (2006) found that nitrification was significantly inhibited during early spring at a centralized piggery waste treatment plant. It was believed that this was due to the overuse of disinfectants in the pig houses. Won and Ra (2001) studied the effect of disinfection on SBR (sequencing batch reactor) units for piggery waste and reported that there was significant effect on nitrification if the disinfectant concentrations exceeded 13 mg/L as cresol. Inhibition started at a concentration of 5 mg/L. However, there is no report available to show how the sludge produced during this period of overdose affected plant growth when the sludge was applied to land. Figure 4.3 describes the fates of organic chemicals such as antibiotics, growth promoters and disinfectants used in a piggery waste treatment scheme when applied to land. The residual organics can be degraded in the soil environment, remained in an adsorbed condition, removed by the plant or washed off by the rain.

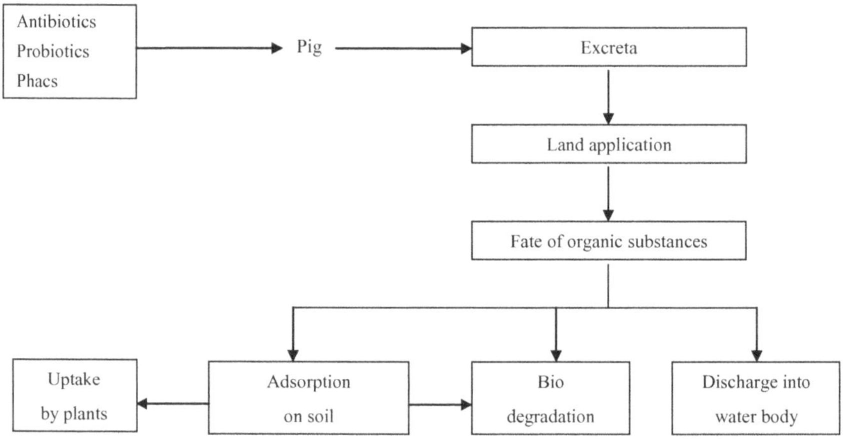

Figure 4.3 Fate of trace organic chemicals applied to land. (Note: Phacs = pharmaceuticals and probiotics = growth promoters)

(3) Pathogens

Figure 4.4 denotes the persistence of selected enteric pathogens in water, wastewater and soil, and on crops. These data are probably based on human excreta, but they could be used to give a general approach on how to handle animal waste. To remove pathogens, a simple applicable technology is storage. Enteric pathogens are killed with storage of about three months while ascaris survives for a year. In Korea, resulting storage requirements ranges from 4 to 9 months for agricultural application (Korean RDA, 2002), but the proposed required storage for municipal sludge was one year in Sweden (Kvarnstrom E *et al.*, 2003).

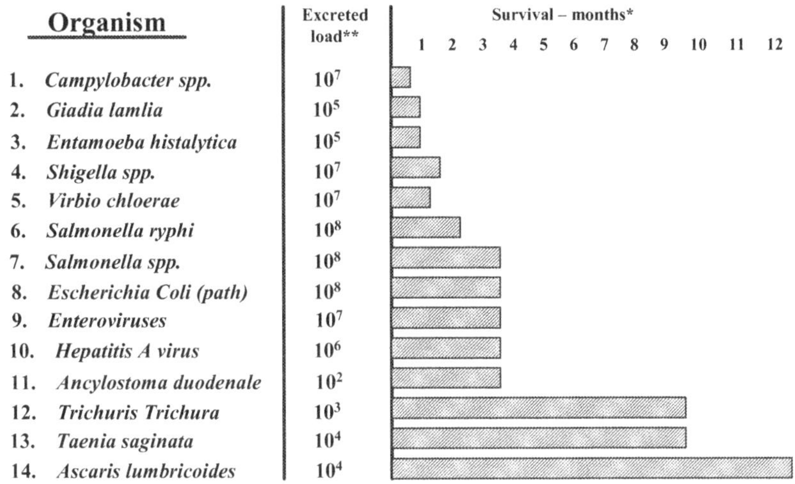

* Estimated average life of infective stage at 20°C - 30°C

** Typical average number of organism/gm feces

Figure 4.4 Persistence of selected enteric pathogens in water, wastewater and soil, and on crops (Shuval, 1991).

Biosolids produced from compost, anaerobic and aerobic digestion may contain organisms and substances which can cause an impact on human and environmental health if they occur above a certain level. These include pathogens, trace elements (especially heavy metals) and synthetic organic compounds. Animal waste is believed to be safer than sewage sludge, since it does not include toxic substances and heavy metals discharged from industries; however, there are some recent animal related health hazards such as pig cholera, mad cow disease and avian influenza.

(4) Harmful Substances Accumulation

A sustainable nutrient management strategy reintegrates nutrients in the environment without accumulating harmful substances above an acceptable level. The combination of heavy metals and nutrients, such as the ratio of Cd/P or Zn/P, is often used to characterize applied waste, similar to chemical fertilizers. It is reported that piggery wastes contain high zinc and copper (Bull, 2005).

Wastes applied to agricultural soils consist of organic and inorganic matter. Organic matter in these wastes differs over a broad range: chemical fertilizer

0%, compost 30%, sewage sludge 50% and manure 90%. Depending on the type of organic matter (easily degradable versus stable), the organic matter will be mineralized partly or completely within a certain time span. Also, the solubility of organic components varies (slightly soluble versus non-soluble). In the long run, most organic matter will decompose and the inorganic matter will partly dissolved. The remaining part or stable matrix has a "diluting effect" on the heavy metal load of the wastes applied on the soils. Table 4.2 shows the total metal load applied which causes their accumulation in the soil. Zn and P concentrations and the anthropogenic loadings of the various fertilizers (mineral P fertilizer, pig manure, sewage sludge and compost) are compared are also shown. As it can be seen, the impact of the "diluting effect" for compost is the highest. Chemical fertilizer has the lowest risk of Zn accumulation. Piggery waste shows the highest Zn load on a basis of P applied among the other fertilizers. Piggery waste contains higher cupper although not shown in the table.

Table 4.2 Dilution impact on the heavy metal accumulation effectiveness of various fertilizers (Lampert, 2003).

Fertilizers	Zn mg/kg DM	OM in %	Anthropogenic. loading (AL) mg Zn/kg DM	P mg/kg DM	mg Zn/kg P	Specific Added AL mg Zn/kg P
Chemical P fertilizer	220	0	193	32,700	6,628	5,900
Pig manure	1,300	90	1,294	26,500	50.189	48,850
Sewage sludge	800	50	772	20,000	40,000	38,575
Compost	200	30	160	4,000	50,000	40,025

Assumptions: Background Zn concentration in the soil: 60 mg Zn/kg DM (dry matter); Decomposition of the organic matter (OM); all 100%; easily soluble inorganic matter: mineral fertilizer 55%, all others 5%.

(5) Antibiotics/ Probiotics /Pharmaceuticals

Figure 4.3 shows the fate of pharmaceutical medicines used in animal farming. Antibiotic residues and increased numbers of antibiotic resistant bacteria have been reported near confined animal feeding operations (CAFOs) and in agricultural watersheds (Campagnolo *et al.*, 2002; Chee-Sanford *et al.*, 2001). In the year of 2000, 92% of swine CAFOs reported using antibiotics in a nationwide survey, and the most commonly administered drugs were tetracycline, tylosin, a macrolide antibiotic (USDA, 2002). Swine CAFOs often use antibiotics for therapeutic needs but mostly for growth-promoting purposes. Many different kinds of antibiotics, probiotics (growth promoters) and pharmaceuticals are available for animal industries. Other countries also use antibiotics. Fortunately, the amount of antibiotics used for nontherapeutic purposes has been gradually reduced.

The reduction of veterinary medicinal products is mainly due to the application of legislated maximum residue levels (MRLs) designated to protect the health of humans who consume food derived from animals (http//europa.eu.int). These MRLs define the maximum threshold for traces of veterinary medicinal products which may be contained in foodstuffs from animals to which these substances have been administered.

Table 4.3 Summary of sorption data to soils for veterinary pharmaceuticals (Tolls, 2001).

Pharmaceuticals	Adsorption coefficient for soil ($K_{d, solid}$ L/kg)	Organic carbon coefficient (K_{oc} L/kg)
Tetracycline	400-1620	
Oxytetracycline	0.3-1030	17-93,300
Enrofloxacin	260-6310	16,500-770,000
Ciprofloxacin	430	61,000
Ofloxacin	310	44,100
FCQA	285	1,100
Enro-CO_2	7.7	
Oxolinic acid	0.3-116	17-4,510
Efrotomycin	8.0-290	580-11,000
Avermectin	7-134	5,300-30,000
Tylosin	8.3-128	550-7,990
Sulfathiazole	4.9	200
Sulfamethazine	0.6-3.1	60-208.3
Metronidazole	0.54-0.67	39-56
Olaquindox	0.69-1.67	46-104
Chloramphenicol	0.2-0.4	

Some of the medicines applied with feed will also be discharged with the excreta. The medicine residuals in excreta are usually easily adsorbed into soil and biologically degraded; however, some of them can be discharged into a water body as shown in Figure 4.3, e.g. antibiotics were found in streams (Kolpin *et al.*, 2002). Table 4.3 gives a summary for the sorption properties of veterinary pharmaceuticals to soils (Tolls, 2001). Table 4.4 shows the estimated concentrations of antibiotics in the manure of beef cattle and swine in a confined feeding operation (Huang *et al.,* 2001). Antibiotics were not found to inhibit nitrification or be toxic to soil fauna. Campos *et al.,* (2001) monitored the toxicity effect of two broad-spectrum antibiotics (chloramphenicol and oxytetracycline) by continuously adding them into a nitrifying culture system. Nitrifiers were present as suspended and attached cultures. No significant change was observed on the surface area covered by the nitrifying biofilm, nor on the nitrification rate when chlorampheicol was applied to the culture between 10 - 250 mg/L. However, nitrification was inhibited at 100 mg/L of oxytetracycline. At 10 mg/L of oxytetracycline, biofilm sloughing was initiated. However, Cho *et al.,* (2006) reported repeated application with a concentration greater than 20 mg/L reduced nitrification and COD removal in a laboratory biological unit operated with diluted piggery waste consisting of 120 mg/L NH_4-N and 500 mg/L soluble COD. The toxic effect of tylosin and oxytetracycline was tested on soil fauna including earthworms, springtails and enchytracids by Baguer *et al.,* (2000). Neither of the antibiotics had any effect at environmentally reasonable concentrations. The compost process can degrade dioxin, but a liquid state biological oxidation treatment process commonly used for pig farms discharges tetracycline.

Amy and Drewes (2005) studied the fate of organics from municipal secondary effluents (EFOM) with a soil/aquifer treatment (SAT) process called a bank filter. Results of the study may be related to piggery waste residual organics:

- Permeability is the most important design parameter of the SAT process which works like a slow sand filter by developing a biologically active film on the top of the filter called a "smutzdeck."
- Removal of DOC (dissolved organic carbon) and DON (dissolved organic nitrogen) increased with depth, but the SUVA (specific UV absorbance ratio of UV_{254}/DOC representing aromatic and aliphatic compounds) remained constant.
- Initially, polysaccharides were removed, then proteins followed by the removal of humic substances.
- Pharmaceuticals were easily removed, but antiepileptic and analgesics were persistently discharged for many years.
- EDCs (endocrine disrupting chemicals) are hydrophobic and are easily removed by biodegradation and adsorption, but can be discharged down stream.
- Chlorinated flame retardants are very stable and difficult to remove.

Table 4.4 Estimated concentrations of antibiotics in the manure of beef cattle and swine from a confined feeding operation (Modified from Hu.ang et al., 2001).

Antibiotic	Class	Antibiotic concentration (beef cattle)	Antibiotic	Class	Antibiotic concentration * (swine)
Chlorotetracycline Sulfamethazine	Tetracycline Sulfonamide	3,973	Chlorotetracycline Sulfathiozole	Tetracycline Sulfonamide	133,000
Tylosine	Macrolide	1,946	Penicillin	B-lactam	
Bacitracin	Polypeptide	1,703	Tylosine	Macrolide	
Oxytetracyline	Tetracyline	1,064	Sulfamethazine	Sulfonamide	89,000
Chlorteracyline	Tetracyline	993	Chlorotetracycline	Tetracyline	37,000
			Bacitracin	Polypeptide	27,000
			Oxytetracycline	Tetracyline	21,000
Erythromycin	Macrolide	730	Lincomycin	Aminoglycoside	11,000
Thiocyanate			Tylosine	Macrolide	11,000
			Oleandomycin	Macrolide	6,000

* ug of antibiotic/kg of manu-e

In contrast with sewage sludge, animal waste is cleaner. If sewage sludge can be applied to an agricultural area, there is no doubt that animal waste can be applied.

Manure generated at CAFOs containing excreted residuals is commonly stored in earthen lagoons for several months before land application. Incomplete degradation of pharmaceuticals can be a contributing factor to the presence of these drugs in waterways. However, during this period, degradation by photo radiation and soil adsorption can take place. Figure 4.5 shows ciprofloxacin mass balance for the Zurich-Werdholzi wastewater treatment plant.

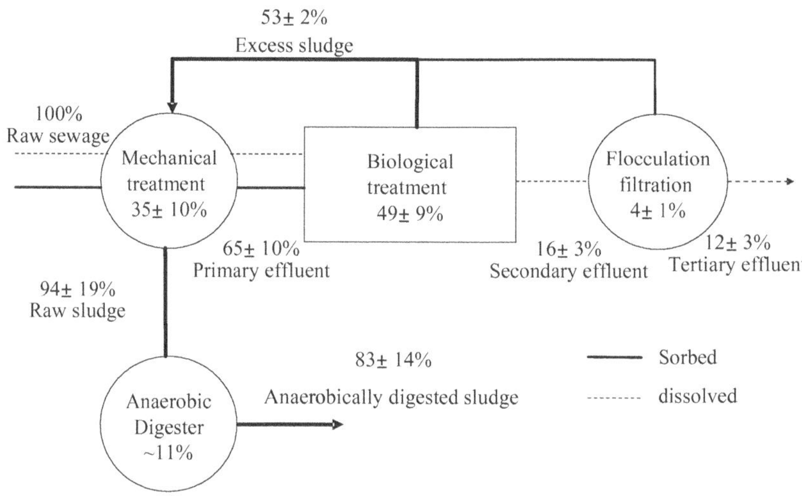

Figure 4.5 Ciprofloxacin mass balance for the Zurich-Werdholzi wastewater treatment plant (Modified from Golet *et al.*, 2003)

Photodegradation with UV resulted in significant changes of tetracycline, olaquindox, penicillin G, spiramycin, streptomycin and tiamulin during test periods of 3 to 7 days (Halling-Sorensen, 2000). Pharmaceuticals, such as tetracycline with greater soil adsorption, exist in surface soil and probably photodegraded since tetracyclines were not detected in the soil below 30 cm or with ground water samples (Hamscher *et al.*, 2002).

After oral administration of tylosin at a dose rate of 110 mg/kg feed, swine were found to excrete up to 40% of the tylosin in the antibiotically potent forms A, B, C and D along with at least 10 other degradates in a range of about

20 mg/L tylosin (Teeter and Meyerhoff, 2003). Tylosin A has average half lives of 4 to 8 days in swine, calf and chicken manure, and 2 to 8 days in aqueous water (De Liguoro *et al.,* 2003; Teeter and Meyerhoff, 2003). A recent study with open and closed lagoon slurries (Kolz *et al.,* 2005) indicated that the majority of tylosin was rapidly sorbed and degraded, with 90% disappearance occurring in less than five days. Both abiotic and biotic degradation were apparent. However, disappearance in anaerobic slurries showed after 24 hours, and a residual pool of tylosin remained through the extended anaerobic incubation. Aerating slurries significantly reduced tylosin residuals, leaving less than 1 % of that added after 12 days. Tylosin B and D were produced during the studies and remained after eight months of anaerobic incubation in a conventional lagoon slurry. This suggests that residual tylosin degradates, with antibiotic activity, may be applied to agricultural fields and contribute to the detection of antibiotic residues near the CAFOs.

4.3 COMPOSTING

Aerobic composting is a method of piggery waste stabilization in which organics are decomposed by microorganisms in the presence of oxygen. Composting results in the production of a stabilized humus-like product that can be used as fertilizer and soil amendment, for erosion control, as a mulch or other soil like products. Aerobic composting is dependent on several operating parameters, which include oxygen availability, moisture content, temperature, C/N ratio and biodegradable part of the organics.

An adequate oxygen supply is necessary for the aerobic microorganisms. If oxygen is not sufficient, anaerobic microorganisms become predominant and produce nuisance odors. If the oxygen supply is too much, it will decrease the temperature and moisture content of the compost pile.

The moisture content must be sufficient to give a favorable environment for microbial growth. If there is too much water, anaerobic conditions can develop. Piggery wastes have a high moisture content, so the addition of a bulking agent is necessary to give the desired moisture content of 60%. The porous nature of the compost mixed with the bulking agent allows for easier air passage to stimulate aerobic activities.

The carbon to nitrogen ratio (optimal 25 to 40) is also important to avoid slow composting (at high C/N ratios) or ammonia volatilization (at low ratios).

A neutral pH is preferred but a range of 5.5 to 8.5 is acceptable. In situations where low pH is a problem (acidity), it may be adjusted with the addition of lime. In hot weather, fermentation causes the mixture to turn acidic, especially for in-vessel compost methods (Haelzle, 2005).

The most common composting techniques include windrow, aerated static pile and in-vessel. Windrow is the least capital-intensive method but requires a large amount of space and labor relative to other techniques. The windrowed piles must be turned periodically to facilitate aerobic conditions. Aerated static pile composting requires mechanical aeration to maintain aerobic conditions. The most common in-vessel technologies are rotating drum and channel approaches (Haelzle, 2005). Composting in channels requires the materials to be turned mechanically or aerated using a forced air system.

In-vessel systems require a higher level of capital investment but systems are more scalable than others. Most channel compost systems require some type of enclosed structure to prevent process control issues, such as moisture build-up from rainfall infiltration. The requirement for an enclosure makes the channel composters less scalable, as the operator must either build the structure to accommodate future expansion or build individual enclosures for each additional channel component. The rotating drum technology is a stand-alone processing solution that uses the rotation of the drums to maintain aerated conditions. The automatic rotation mechanism makes the rotating drums relatively easy to use with a relatively low capital cost. It also requires slightly more care in the initial blending of the compost to ensure optimum performance of the bacterial action in the drum.

Pre-mixed materials are loaded into the drum with a loader/hopper or a conveyer system (Haelzle, 2005). The material enters the front of the drum and is turned toward the back of the drum over a three to four day period. Temperatures in the drum ensure the destruction of pathogens and weed seeds over the initial composting period. Materials comes out of the drum in a relatively clean dry state and can be used relatively soon as a bedding material.

If the operator desires to move the composted materials to a market for profit, the materials must typically undergo 15 to 25 days of curing in windrows, depending on the rules of the targeted market. Even where this is the case, the area required for in-vessel compost systems is less than that for windrow or static pile systems due to the on-ground residency time required to bring windrow or static pile composted materials to commercial grade compost.

The selection of a bulking agent or conditioner is largely depended on the local availability. Rice husk, straw and wood waste (e.g., saw dust, wood chips) are commonly used and corn stems or cobs also can be used. These materials absorb moisture, give porosity for aeration of the mixture and are degraded with the piggery waste during the composting process.

All kinds of agricultural and natural products used for bulking agents are believed to be biodegradable. Some products are easily decomposed and other products take a long time to be degraded depending on given environmental conditions. Rice straw and husk are more degradable bulking agents than tree products such as planer waste and degraded tree residues. Saw dust is a commonly used as a bulking agent, and its degradation rate is greater than the planer residue, since its size is smaller. The degradability of bulking agents within a limited time would be increased as the size of bulking agent is smaller. Figure 4.6 shows the measured relative degradation rates of various bulking agents at laboratory condition.

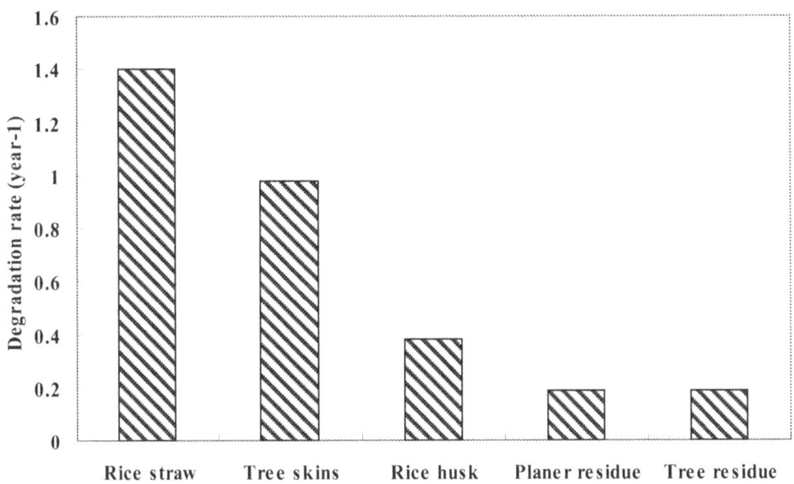

Figure 4.6 Relative degradation rates of various bulking agents under aerobic condition with a moisture content of 60% at 35°C.

During composting, aerobic degradation releases heat within the compost. In the temperature range of 50 to 70°C, pathogens are killed and water evaporates. Temperature-time requirements for pathogens kills are 55°C for 2 weeks, 60 °C for 6 days, 65°C for 3 days and 70°C for 1 day, as shown in Table 4.5 (Kvarnstrom *et al.,* 2003). Temperature control can be done by regulating the air supply and/or by the frequency and degree of mixing. A reduced hot air supply and more frequent mixing will reduce the temperature. Water vapor is also removed by aeration. Animal farms in Korea generally use a channel composting system with a travelling mixing device without force aeration in a vinyl house or building. The mixers are mounted on rails which move from the front to the end of the compost pile to provide oxygen and mixing. Sun light heats the composts to higher temperatures in vinyl housing during the summer. Figure 4.7 shows temperature and moisture profiles in a channel composting plant processing piggery wastes (Park S W, 2001).

Table 4.5 Proposed treatment processes for sewage sludge, divided into categories A, B and C in Sweden. All values are given as minimum values (Kvarnstrom *et al.,* 2003).

Category	Treatment process	Parameters to be fulfilled
A	Thermal drying	Temperature: ≥ 90°C, Exposure time: 10 minutes, Moisture: < 10%
A	Pasteurization	Temperature: ≥ 70°C, Exposure time: 60 minutes
A	Thermophilic anaerobic digestion	Temperature: ≥ 55°C, Exposure time[a]: 6 hours Mean retention time[b]: 7days at ≥ 55 °C
A	Liquid composting	Temperature: ≥ 55°C, Exposure time: 10 hours Mean retention time[b]: 7 days at ≥ 55°C
A	Thermophilic aerobic stabilization (in vessel composting)	Temperature – time requirements: 55 °C for 2 weeks, 60°C for 6 days; 65°C for 3 days and 70°C for 1 day
A	Conditioning with lime (quick lime)	pH: ≥ 12 and, Temperature: ≥ 55°C, Time: 4 hours
B	Thermophilic aerobic stabilization in windrows or piles (open composting)	Temperature – time requirements[c]: 55 °C for 2 weeks; 60°C for 6 days; 65°C for 3 days and or 70°C for 1 day
B	Conditioning with lime (slaked lime)	pH: ≥ 12 Time: 3 months
C	Treatment and drying in reed beds[d]	1 year without addition of new sludge
C	Storage	1 year without addition of new sludge

[a] Exposure time is defined as the time when no sludge is added or withdrawn
[b] The mean retention time is calculated to include at least 95% of the material
[c] The temperature – time requirement should be met during three periods, the material being turned over between each period
[d] Methods involving dewatering and possibly biological degradation

Figure 4.7 Temperature and moisture profile in a channel composting plant.

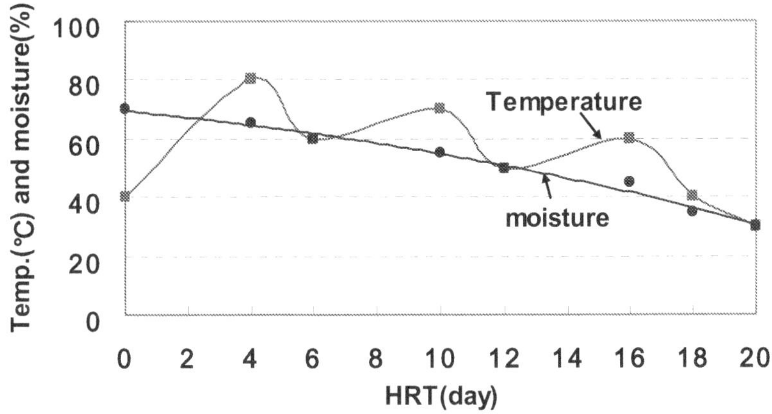

4.3.1 Retention Time

The minimum composting time varies with kind of amendment or bulking agent used for piggery waste. It is generally 20 to 30 days excluding a curing period of two to three months. The temperature-residence time requirement to kill pathogens must be met and organics must be stabilized. Stabilization is defined as the reduction of the decomposition properties of biowaste to such an extent that offensive odors are minimized and that either the respiration activity after 4 days (AT4) is below 10 mg O_2/g-dry matter or the dynamic respiration index is below 1,000 mg O_2/kg VS/hr (Spinosa *et al.*, 2006).

4.3.2 Aeration Requirement

Aeration is required to supply oxygen for microbial degradation, to remove heat accumulation and to provide drying. Aeration can be provided by forced aeration or turning the compost. Assuming a mixture composition of piggery waste and bulking agent of $C_{10}H_{19}O_3N$,

$$C_{10}H_{19}O_3N + 12.5\ O_2 = 10\ CO_2 + 8\ H_2O + NH_3 \quad (4.1)$$

(201) (400)

The oxygen demand is estimated to be 2 kgO_2/kg VS. Most of the ammonia will not be oxidized because of temperatures exceeding 35°C. Some will be used for cell synthesis, but the largest fraction will probably be lost from the compost by volatilization. Therefore, oxygen demand for nitrification is not generally considered. The stoichiometric demand for oxygen varies from a low of about 1 kg O_2/kg organic for highly oxygenated wastes such as cellulose to a high of 4 kg O_2/kg organic for saturated hydrocarbons. Amendments such as sawdust are largely cellulosic in nature, with a degradation that can require 1.2 kgO_2/kg VS (Lue-Hing C, 1992).

4.3.3 Applicable Standards

Tables 4.6 and 4.7 show European Eco-Label standards for compost materials which includes heavy metals, nitrogen, phosphorus, pathogens and others constituents. Piggery wastes have higher copper and zinc contents. The limits of copper and zinc vary greatly with countries; 90 to 1,750 mg/kg for Cu and 400 to 4,000 mg/kg for Zn.

Table 4.6 Europe eco-label standards applicable to composts.

Tested traits	Limits as determined by Test Methods 86/278/EEC
Heavy metals	See Table 4.7
Special metals	If contains industrial or municipal wastes, then test for: Mo, Se, As, F
Constituents	Organic Matter > 20%: Moisture <75%: Total-N less than 2% TS
N - P_2O_5-K_2O Application limits	Application rates shall specify not more than: 17g/m^2 N – 6 g/m^2 P_2O_5–12 g/m^2 K_2O
Pathogens	Salmonella nondetect in 25 g *E.coli* < 1000 MPN/g
Other:	Contains no offensive odors; No glass, wire or other fragments; No unacceptable weed seeds
Declarations:	Must describe recommended use and application rates; All feedstocks > 10% must be reported; Nutrients, organic matter and metals must be reported; No phytotoxic effects

Source: Official Journal of the European Community (OJ, 1998).
Note: Conversion from P_2O_5 to P_2: $2P/P_2O_5$ = 31 x 2/ (31 x 2 + 16 x 5) = 62/142 = 0.44
Conversion from K_2O to K_2: $2K/K_2O$ = 39 x 2/ (39 x 2 +16) = 78/94 = 0.83

Table 4.7 Heavy metals limits (mg/kg) for European countries[a] which do have compost rules.

Element	A	A[b] Class 2[c]	B Agr[c]	B Park[c]	CH	DK	F	D	I	NL	NL	SP	C A,AA[c]
Arsenic	-	-	-	-	-	25	-	-	10	25	15	-	13
Boron	100	-	-	-	-	-	-	-	-	-	-	-	-
Cadmium	4	1	5	5	3	12	8	1.5	1.5	2	1	40	3
Chromium	150	70	150	200	150	-	-	100	100	200	70	750	210
Cobalt	-	-	10	20	25	-	-	-	-	-	-	-	34
Copper	400	100	100	500	150	-	-	100	300	300	90	1,750	100
Lead	500	150	600	1,000	150	120	800	150	140	200	120	1,200	150
Mercury	4	1	5	5	3	1.2	8	1.0	1.5	2	0.7	25	0.8
Nickel	100	60	50	100	50	45	200	50	50	50	20	400	62
Selenium	-	-	-	-	-	-	-	-	-	-	-	-	2
Zinc	1,000	400	1,000	1,500	500	-	-	500	500	900	280	4,000	500

a. Country code: A Austria; B Belgium; C Canada; DK Denmark; F France; D Germany; I Italy; NL Netherlands; SP Spain; CH Switzerland
b. Calculated on 30% organic matter basis
c. Notes: Class-2 for A; for B, Agr =Agricultural use; Park= Horticultural use ; and classes A, AA for C.

4.4 ANAEROBIC TREATMENT

Like compost, stabilization of organics is required with anaerobic and/or aerobic degradation for liquid compost or for the production of bioliquid for agricultural purpose. During stabilization, pathogens are removed and odor problems are minimized. Larger solids will be hydrolyzed to volatile fatty acids which will prevent clogging when applied to the soil.

4.4.1 Anaerobic Treatment Fundamentals

Complex organic matter is converted by hydrolysis to simple, soluble organic matter. The soluble organic matter is converted by acid forming bacteria into volatile fatty acids (VFA), which are broken down into CH_4 and CO_2 by methane forming bacteria in the absence of oxygen. The hydrolysis and acid formation reactions comprise the fermentation process. The amount of methane gas production can be easily computed from following equation:

$$2\,C + 2\,H_2O \rightarrow CH_4 + CO_2 + \text{bacterial mass} \quad (4.2)$$

$$2\,C + 2\,O_2 \rightarrow CO_2 \quad (4.3)$$

A total of two moles of carbon (equivalent to 64 g COD) produces one mole of CH4, which is equivalent to 22.5 L at standard condition; or 0.35 m^3 CH_4 is produced from one kg COD removed. During this reaction, about 10% of the COD is used for the formation of bacterial cells resulting in an actual CH_4 gas production of 0.32 m^3/kg COD removed.

Biogas is also produced from endogenous respiration as below:

$$C_5H_7O_2N + 4H_2O \rightarrow 1\ 1/2\ CO_2 + 2\ 1/2\ CH_4 + NH_4HCO_3 \quad (4.4)$$

This reaction can increase CH_4 gas production with higher MLSS (mixed liquor suspended solids) operation in the anaerobic digester, but the overall gas production of 0.32 m^3 CH_4/kg COD removed will be almost consistently maintained. The gas produced from endogenous respiration has a higher CH_4 content of 83% with a higher pH due to alkalinity production.

The CH_4 content will depend on the alkalinity and pH of the system. Generally, higher pH and higher alkalinity produce a higher CH_4 content. Piggery waste in Korea generally shows a higher pH and alkalinity, and the CO_2 content is very small or negligible (Choi *et al.*, 2005).

COD or VS (volatile solids) are frequently used to evaluate the performance of digester, e.g., m^3 gas/kg COD removed or m^3 gas/kg VS removed. The purpose of anaerobic digestion of municipal sludge is sludge volume reduction. However, the performances of anaerobic digesters for piggery waste management on a COD basis is more reasonable since waste stabilization is the main purpose, providing there is a relationship between COD and VS.

There are many different shapes of digesters. Complete mix, plug flow, biofilm, UASB and others. In addition, there are unheated and heated digesters. A discussion of the shape of digesters and their mechanisms can be found in most wastewater treatment text books and other related reference materials.

4.4.2 Bioliquid from Anaerobic Storage

Storage permits solid settling and anaerobic hydrolysis of solids into simple forms of organic acids which are easily removed by microorganisms in the liquid and soil. Furthermore, fermented organics including aldehydes and alcohols can be converted into CH_4 and CO_2. Figure 4.8 denotes how COD and/or BOD of nightsoil and piggery waste vary with storage periods under different environmental conditions. Generally, about 30 to 80% of the COD or BOD was removed with storage, but the variances largely originated from the settleability and digestability of the nightsoil. The solids were settled as well as floated. For the 60 to 70% COD removal, only about 10% of the total COD removal was estimated to be removed by conversion to CH_4 (Chung, 1981). In other words, 90% of the total COD removal originated from settling. Soluble COD removal was not apparent. In order to increase its removal efficiencies, the sludge which settled down must be removed periodically as discussed with the anaerobic baffled reactor (See Chapter 3).

An anaerobic lagoon is a common type of storage facility, but it requires a large volume. After storage in an anaerobic lagoon, the lagoon overflow is land-applied by spray to grow crops for animal food. Also, a series of facultative or aerobic lagoons may be installed for water and nutrient reuse. Overflow with storms, leakage from the lagoon bottom, offensive odors, ammonium volatilization and public health are common problems with anaerobic lagoon operation.

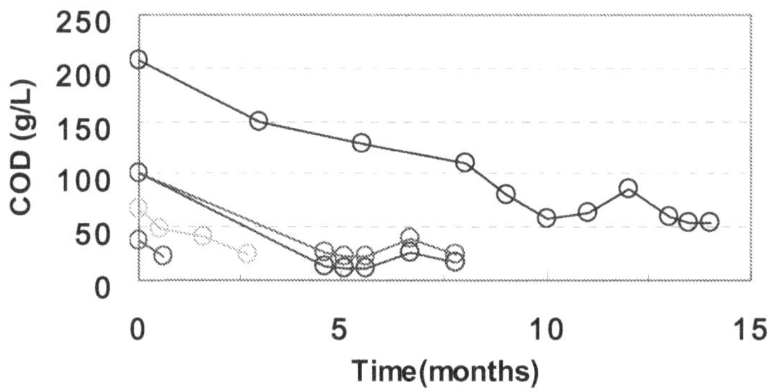

(a) COD changes in nightsoil

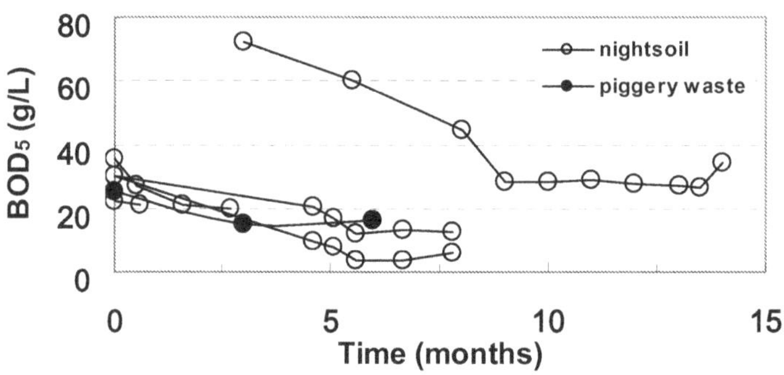

(b) BOD changes in nightsoil and piggery waste

Figure 4.8 COD and BOD changes with storage time from different initial concentrations during anaerobic storage. Nightsoil is a mixture of human feces and urine. The characteristics of nightsoil and piggery waste are fairly similar, each with high organics and nitrogen contents as shown in this figure. Piggery waste data are from Korean RDA (2002).

4.4.3 Lagoons

Odor problems are severely limiting the continued permitting and use of lagoon systems in North Carolina where there were about 4,500 active and 1,700 inactive lagoons in 2002. Anaerobic lagoons began to be used extensively for piggery wastewater treatment in the mild climate of the southeastern region of the Unites States around 1965. Discharge from these lagoons was no longer allowed as a result of federal legislation passed in 1972. Therefore, lagoon effluent was used for irrigation using fertilizer rates based on nitrogen. These lagoons provided up to 90% removal of nitrogen and storage so that wastewater could be applied conveniently to less acreage than was required for the discharged waste. The benefits of this type of system are verified by the fact that more than 4,000 piggery waste lagoons were in operation in North Carolina by 1999. (Humenik, 2005).

Early lagoon loading recommendations were made on the basis of pig population per surface area: 3,750 pigs/ha of lagoon (40 g BOD/m^2, 23 g BOD/m^3) in the northern US to 15,000 pigs/ha of lagoon (168 g BOD/m^2, 90 g/m^3) in the southern US (Eby, 1964; SCS, 1970). This is equivalent to 1 to 24 m^3 of lagoon volume per 45 kg hog (Humenik and Overcash, 1976) Below the load of 9.2 to 18.4 m^3/45 kg hog, odor was not manure-like, nor was an odor always detectable. These results led to lagoon loading values for North Carolina of 0.028 to 0.083 m^3/45 kg pig, with 0.056 m^3 (2 ft^3/100 lb pig) being the general recommendation and 3 ft^3/100 lb pig being recommended where more positive odor control was desired. These rates are still recommended for North Carolina (NRCS, 1989).

4.4.4 Covered and Heated Anaerobic Digestion

Covered and heated anaerobic digesters are common in Europe. This technology is well developed and practical aspects of its application to animal waste management are introduced in Chapter 7. The USA uses a series of lagoon systems for recycling nutrient and water from their management systems. These are mostly uncovered resulting in persistent odor and overflow problems from rainfall. Thus, there have been lagoon modifications to correct these problems. Improvements for simple lagoon operation are shown in Table 4.8. Covered lagoons reduced odor, greenhouse gas emissions and improved water quality. Heating the digester and providing aeration were also effective.

The ambient temperature covered lagoon digesters are most successful in warmer climates. A properly sized lagoon receives dilute manure with less than 2% TS. Gas production varies seasonally and is useable for electricity production, or is burned in a flare for odor control. Complete mix tank digesters are used to treat waste with 3 to 10% TS. The units are usually heated with hot water from an engine to 35ºC and mixed to maintain a high level of bacterial activity. The HRT is greater than 12 days. The heated mixed covered lagoon requires 1/3 or less of the volume required for ambient temperature lagoons to perform the same function. Roos (2003) compares the environmental effectiveness of various treatment options for odor control, GHG reduction, water quality protection and costs (Table 4.8). Further, it was found that multiple cells were more efficient for removing *fecal coliform* and *Salmonella*.

Table 4.8 Environmental effectiveness of manure management options (Roos, 2003).

Options	Odor control	Greenhouse gas reduction	Water quality protection	Cost range (per 500 kg live weight)[++]
Covered lagoon digesters with open storage ponds	E	H	G	\$150-400
Heated digesters (i.e., complete mix and plug flow) with open storage tanks	E	H	G	\$200-400
Aerated lagoons with open storage ponds[+]	G-E	H	F-G	\$200-450
Separate treatment lagoons and storage ponds 2-cell systems)	F-G	L	G	\$200-400
Combined treatment lagoons and storage ponds	P-G	L	F-G	\$200-400
Storage ponds and tanks	P-F	M-L	P-F	\$50-500

Key: **P** = poor, **F** =f air, **G** = good, **E** = excellant, **M** = medium, **H** = high
[+] For aerated lagoon energy requirements, add an additional \$35.50 per 500 kg/year.
[++] Cost ranges do not include annual operation and maintenance {O&M} costs.

Complete mix type reactors with 2.5 to 10% TS must have retention times greater than 17 days with temperatures between 35 to 40ºC. Biofilm digesters with less than 5% solids can have retention times from 1 to 6 days depending on the waste biodegradability for temperatures from 15 to 40ºC (Conservation Practice Standard, 2003). Covered unheated digesters will have different retention times depending on ambient temperatures with a maximum loading of 12 lb VS/1,000 ft^3/d (192 g VS/m^3/d) in Florida and 5 lb VS/ 1,000 ft^3/d (80 g VS/m^3/d) in the northern US.

4.5 THERMOPHILIC AEROBIC DIGESTION

Unlike compost, aerobic bioliquid (or biofertilizer or aerobic digestion) systems do not require bulking agents to supply oxygen to microbes.

4.5.1 COD and Nitrogen Removal Efficiencies

Tables 4.9 and 4.10 show operating conditions for full and laboratory scale aerobic digestion systems, respectively. The heat released from aerobic degradation can increase the reactor temperature. However, the heat can also be lost with the effluent and through the tank walls. The loss will be greater in smaller tanks. Thus, the increased reactor temperature phenomenon results in little temperature increase in laboratory reactors. Therefore, the laboratory studies for aerobic digestion are usually conducted at a controlled temperature.

Figure 4.9 presents removal efficiencies of COD and nitrogen for various organic loads. As organic loads are increased to 10 kg COD/m^3/d at 60ºC, COD removal efficiencies are decreased to 60 %, but nitrogen removal efficiencies remained unexpectedly at 80 %. Actually, nitrification is inhibited at temperatures greater than 35ºC and the removal is assumed to be the result of ammonia stripping. However, there is a possibility that microbial reactions had some effect on nitrogen removal at the higher temperatures, besides stripping. Yi *et al.,* (2003) conducted a high temperature study at 60ºC with exhaust gas analysis and observed N_2O evolution, which is a biological end product. But, it was not mentioned how much biological reaction contributed to its total removal efficiency. In addition, the COD and nitrogen removal efficiencies seem to vary with the feed type and operating conditions, since the study with nightsoil at 45ºC showed less COD and nitrogen removal efficiencies in comparison to the results with piggery waste at 60ºC.

Table 4.9 Full-scale operation of thermophilic aerobic digestion for liquid fertilizer.

Wastewater type	Reactor volume (m^3)	Retention time (days)	COD load (kg/m3/d)	Reactor temperature (°C)	Input pH	Reactor pH	COD removal (%)	N removal (%)	Sources
Cattle slurry	10	4–5	8.3	45–53	-	8.0			Skjelhaugen O J (1991)
Cattle slurry	10	4–7	3.7	48–55	6.5	7.5			Skjelhaugen O J and Donantoni L (1998)
Septic sludge + cattle slurry	10	7–10		50–60	6.8	7–9			Skjelhaugen *et al.* (1992)
Black-water + food watse	10	7		52–56	5.1	7.9			Skjelhaugen and Saethe (1994)
Fish farming sludge	10	7 (batch)		63	6.2	8.7			Skjelhaugen O J and Saethe T (1994)
Pig urine	25	28 10	1.13 2.0	40 41	8.3 8.4	8.3 8.0	96 96	98 97	Eum (2006)

Table 4.10 Lab-scale operation of thermophilic aerobic digestion for liquid fertilizer under controlled temperatures.

Wastewater type	Reactor volume (m^3)	Retention time (days)	COD load ($kg/m^3/d$)	Reactor temperature (°C)	Input pH	Reactor pH	COD removal (%)	N Removal (%)	Sources
Pig slurry	10	3	11	60	7.2	9.5	62	80	Yi *et al.*, (2003)
		1	30		7.3	9.5	58	60	
		0.5	64		7.4	9.6	35	38	
Pig slurry	10	5	2	50	7.8	9.2	65	85	An (2002)
		7	3.5		8.2	9.5	59	79	
Pig urine	60	20	5.1	25	8.6	7.8	92	78	Lee J-H (2006)
	30	10	3.5	35		8.1	90	80	
Nightsoil	75	20	1.9	36/35	7.2	8.9	66	97	Kim YC, Choi E (1991)
		10	5.4	45/43	7.2	8.5	47	24	
		7	6.7	45/43	8.2	8.5	38	6.4	
		5	8.5	44/42	8.2	8.4	32	6.4	

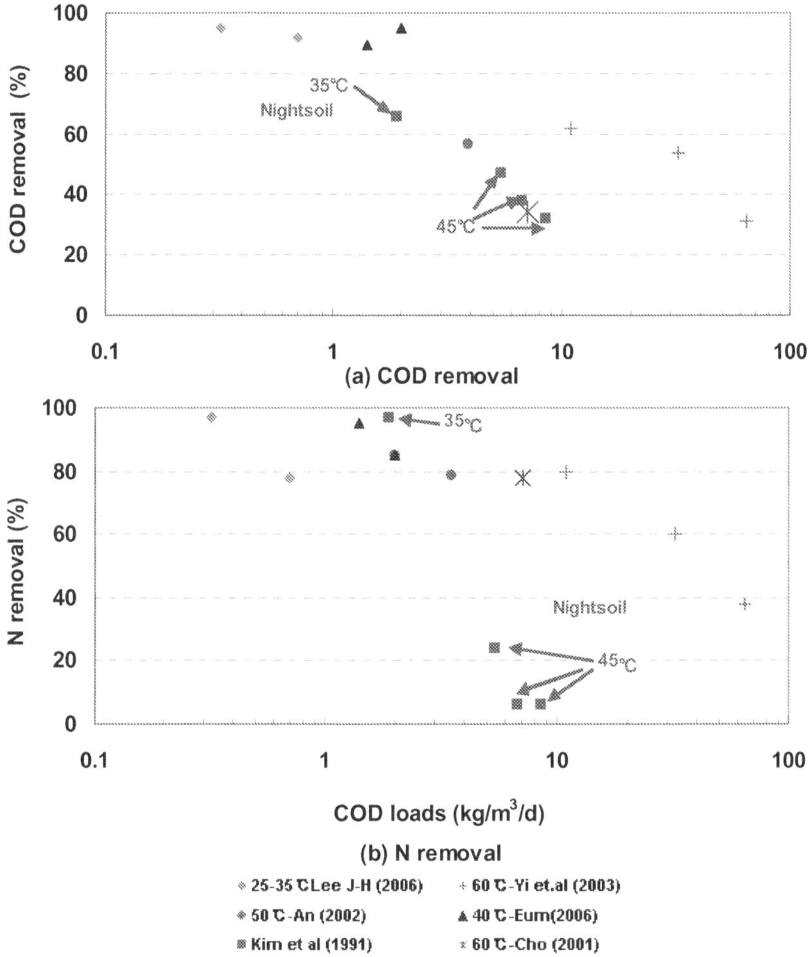

Figure 4.9 Removal efficiencies of thermophilic aerobic digestion treating piggery wastes and nightsoil.

4.5.2 Small Installations

Figure 4.10 illustrates an aerobic liquid compost system for piggery wastes, which includes a fermentor, an auto thermal aerobic digester and settling chambers. The feces and urine are introduced into the fermentor, where the digested liquor is sprayed on top of the fermentor surface to control foaming. The temperature of the fermentor and digester is maintained at about 45 °C. The digester overflow is conveyed to the settling chambers to remove settleable solids. The effluent from the setting chambers is stored. The effluent is recycled back for flushing and the excess effluent can be applied as a fertilizer.

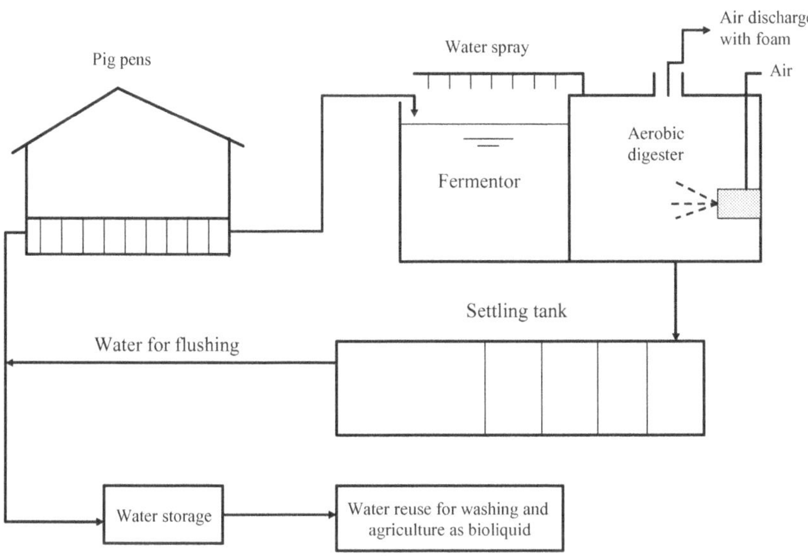

Figure 4.10 Aerobic liquid compost system

It is recommended that the sludge settled in the fermentor be removed every three months and that special microbial seeds (not shown in the figure) be added to the digester for successful operation. The distributor of this process claims that about 70 to 80 % of the water will be evaporated during the processing of the waste.

A proposed design for this system requires about a 15 day HRT for the fermentor and a 15 day HRT for the aerobic digester with an organic load of 2 kg BOD/m^3/d. The submerged turbine aeration system is sized at 1.3 kg BOD/kwh/d. The design temperatures of the fermentor and the digester are 40 and 45°C, respectively.

4.5.3 Expected Reactor Temperatures

Figure 4.11 shows the process streams considered in energy input and output for a thermophilic aerobic digester system. It was assumed that 3,600 Kcal of heat is released from every kg of COD removed (Jewell and Kabrick, 1980) and the heat losses through surface evaporation and the tank walls are 540 Kcal/kg water and 0.4 Kcal/m^2/hr, respectively (Tilati and Stenstrom, 1990). The assumed input air temperatures to the blower are 60°C during the summer and 45°C during the winter. The aeration requirement is computed based on oxygen transfer efficiencies (OTE) as proposed by Surucu *et al.*,(1976). The influent COD and nitrogen concentrations during winter are assumed to be 100 and five g/L, respectively. During summer, the influent COD and nitrogen concentrations are reduced to 56 and 3 g/L, respectively, due to increased flow rates. The influent flow rate is assumed to be 10 m^3/d during winter. Figure 4.12 shows the expected reactor temperatures at different COD loads with different COD removal efficiencies, which can be obtained from Figure 4.9. The expected COD loads to reach 60°C are 3 kg/m^3/d during the summer and 7 kg/m^3/d during the winter. This computational method for the expected reactor temperatures is basically the same as the method used by Lapara and Alleman (1999), except their COD concentrations were lower.

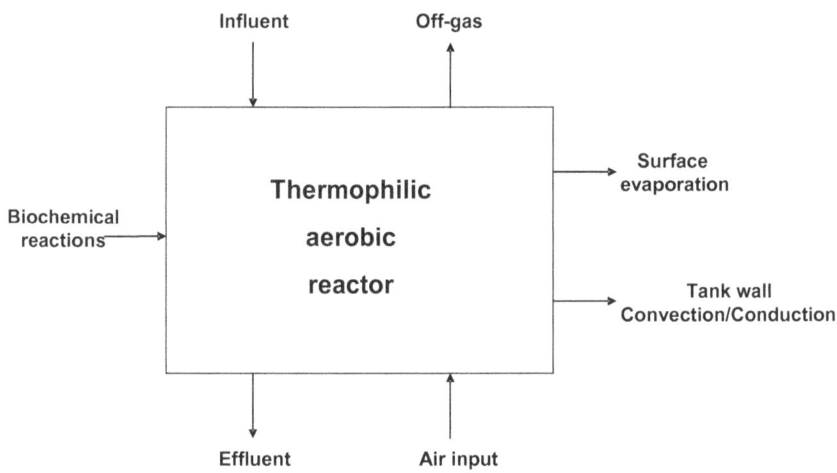

Figure 4.11 Energy balance components for a thermophilic aerobic reactor.

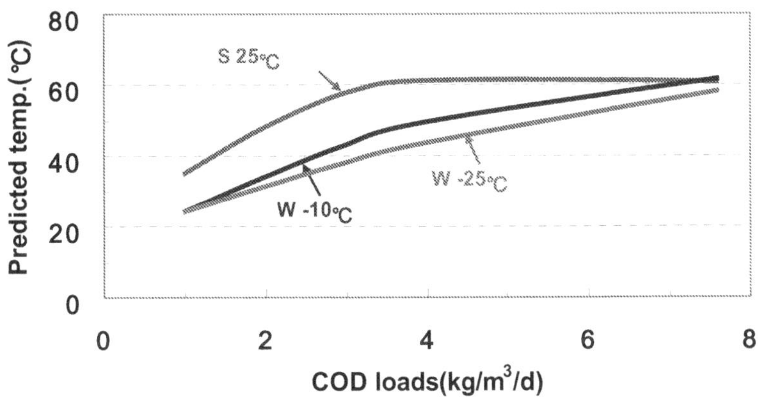

Figure 4.12 Expected reactor temperatures for influent COD loads. (Temperatures in the figure denote air temperatures; S = summer at 25°C, W = winter at -10°C and -25°C, but the influent waste temperature is assumed to be 5°C during the winter).

Figure 4.13 shows the expected reactor temperatures at ambient temperatures of -10°C and 30°C with different oxygen transfer efficiencies (OTE) in full scale operating systems (Eum, 2006). It shows that OTEs are more likely to be 10 to 15% during the winter and 5% during the summer based on actual operating temperatures of full scale aerobic digesters operated at lower loads.

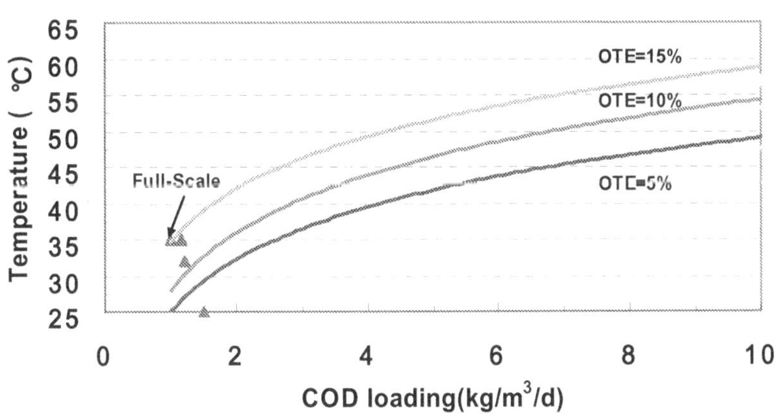

(a) Winter operation at -10°C

Figure 4.13 Expected reactor temperatures at various organic loads and OTE for full scale aerobic digesters.

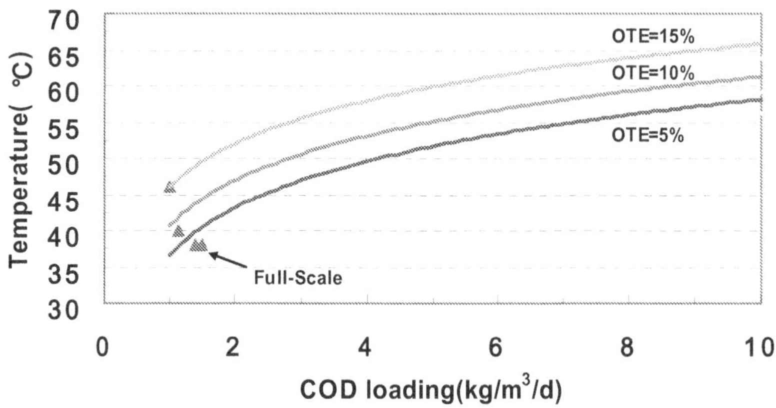

(b) Summer operation at 30°C

Figure 4.13 (continued) Expected reactor temperatures at various organic loads and OTE for full scale aerobic digesters.

4.6 DUCKWEED AND FISH PONDS

Fish growth is limited by the availability of their natural food sources, commonly zooplankton, pelagic algae or benthic animals, such as certain crustaceans and others. Tilapia species filter feed directly on phytoplankton, which makes higher production possible. Photosynthetic production can be increased by fertilizing the pond with artificial fertilizer mixtures containing potash, phosphorus, nitrogen and microelements. Because most fishes are carnivorous, they occupy a higher place in the tropic chain and therefore only a tiny fraction of primary photosynthetical production (typically 1%) will be converted into harvestable fish. As a result, without additional feeding the fish harvest will not exceed 200 kilograms of fish per hectare per year, equivalent to 1% of the photosynthetic production (http://en.wikipedia.org/wiki/Fish-farming).

The other point of concern is the risk of algal blooms, which is called as eutrophication. When temperatures, nutrient supply and available sunlight are optimal for algal growth, algae multiply their biomass at an exponential rate, eventually leading to an exhaustion of available nutrients and a subsequent die-off. The decaying algal biomass will deplete the oxygen in the pond water and lead to massive loss of fish.

In order to tap all available food sources in the pond, the aquaculturist will choose fish species which occupy different places in the pond ecosystem, e.g., a filter algae feeder such as tilapia, a benthic feeder such as carp or catfish and a zooplankton feeder (various carps) or submerged weeds feeder such as grass carp.

For wastewater treatment, oxidation ponds are more popular and have a long history. The ponds typically consist of algae and ordinary heterotrophic organisms (OHO) without fish. Algae produces oxygen through photosynthesis utilizing inorganic substances where the limiting nutrient is phosphorus, and oxygen is utilized by OHO for decomposing organic wastes. In order to keep the pond system in an aerobic condition, the pond surface area must be large enough for the algae to grow. If the pond is not large enough to produce the required oxygen for the degradation of the incoming wastes, the pond will become septic, producing an odor and failure of the system. Alaerts *et al.,* (1996) conducted performance analysis of a full scale duckweed covered sewage lagoon in Bangladesh where the water temperature varies 17 to 30°C.

The use of duckweed and fish farming systems for animal waste management is a common practice. Aquatic carp, fed by duckweek, is the common fish used in this system. For example, the author visited in 2004 a farm located nearby Changchun in the northeastern part of China. The farm collects wastes from pigs, cows and chickens, and puts them into an underground anaerobic digester where no mixing and heating is provided. Biogas collected from digester is used for cooking and heating the house. The digester overflow and sludge are discharged into a duckweed pond. The duckweed is used for fish farming as shown in Figure 4.14. The effluent from the fish pond is discharged into rice paddies.

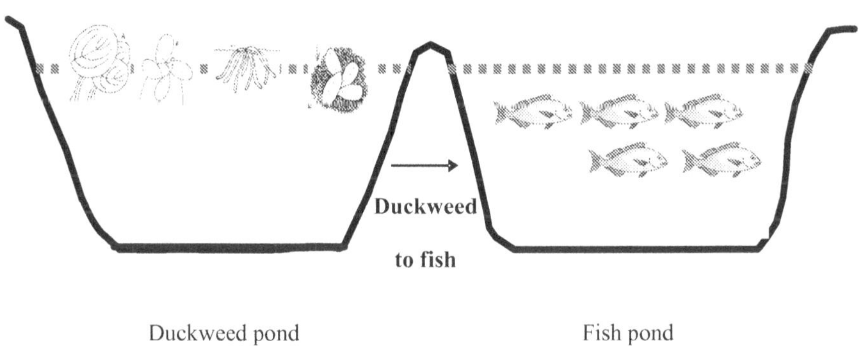

Figure 4.14 Arrangement of duckweed and fish ponds.

Duckweed are small floating and flowering aquatic plants found world-wide and often seen growing in thick, blanket-like mats on still, nutrient-rich fresh and brackish waters. They are monocotyledons belonging to the botanical family *Lemnaceae* and are classified as higher plants, or macrophytes. They are often mistaken for algae. The family consists of five genera as shown in Figure 4.15 among which about 40 species have been identified so far. They share a very simple structure without leaf or stem. The entire plant consists of a soft flat ovoid frond with a size of 2 to 20 mm.

Duckweed grows in nutrient rich waters in a uniform, single layer blankets. Duckweed can grow fast and their biomass productivity can be increased by agronomic management, fertilization and an optimal harvesting schedule (Skillicorn *et al.,* 1993). Duckweed is easy to harvest and has a high protein content (20 to 35% of dry weight).

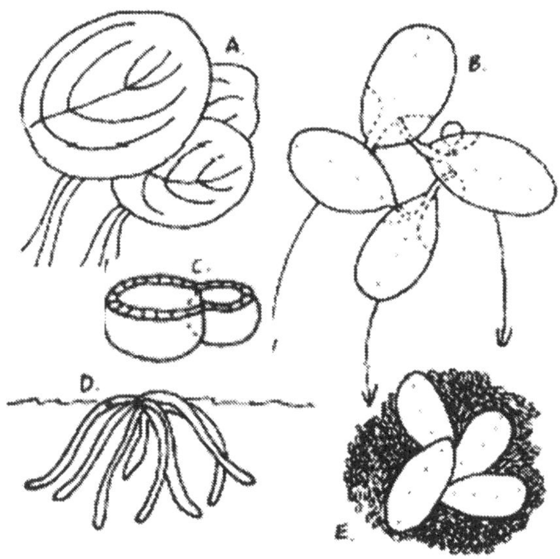

Figure 4.15 Duckweed Genera: A. *Spirodela*, B. *Lemma*, C. *Wolffia*, D. *Wolfiella* and E. *Lemna* with *Wolffia*.

There are many duckweed ponds with fish farming in China and India. Animal and human wastes are fed to the first pond for duckweed growth where they are fed in the next pond for fish growth. The animal management system is known as pisciculture. In India, duckweed ponds operate with 8 to 10 days of

HRT and the fish ponds yield about 10 tons of fish/ha/year (Sulabh International, 2005).

The following is a summary from a book by Skillicorn *et al.,* (1993), who carried out an intensive study with duckweed and fish farming in Bangladesh:

> Contrary to oxidation ponds, nitrogen is a major limiting nutrient in duckweed systems followed by phosphorus with a N/P ratio of 5:1. Sunlight and temperature are important environmental factors for the growth of duckweed. Duckweed can grow in the pH range of 5 to 9, but the optimum pH range is 6.5 to 7.5. The maximum production rate of duckweed is 4 tons/ha/d, but normal production rate is between 0.5 to 1.5 tons/ha/d with a nitrogen feeding rate of 4.5 kg NH_4-N/ha/d. Duckweed is a food for fish and animals. About 110 tons of duckweed produces 14 tons of fish. The major fish is carp which may eat as much as their body own weight daily, but they do not eat during the winter. The minimum DO level for fish is 4 mg/L, which requires close attention at dawn when DO is mostly depleted. Asian carp production is normally 2 to 8 tons/ha/year but can be increased to 10 to 15 tons/ha/year with well managed fish ponds. Overall, the surface area of a fish pond is computed to be about 4 to 5 times the area of the duckweed pond. In order to get 10 tons of fish/ha/year, 80 tons duckweed/ha/yr is required, which is equivalent to 210 kg duckweed/ha/day. The required NH_4-N load is to be about 1 kg/ha/d. If a pig produces 30 g N/d/head, about 30 pigs will be required for 10 tons fish/ha/year if all the nitrogen is used for duckweed production.

Haustein *et al* (1987) studied the pathogen potential of duckweed grown from sewage and fed to chickens in Lima, Peru with the following general conclusions:

> Pathogen reduction in any pond system relies on two simple mechanisms: sedimentation and die-off. Parasites and parasite ova precipitate with other suspended solids and are trapped in the bottom sediment. Other pathogens, suspended in water, simply die as a function of time and temperature. A sufficient detention time must be provided to ensure die-off of pathogens adequate to meet effluent discharge or reuse standards. As with any organic, surface area enhancing material introduced into wastewater, duckweed plants marginally concentrate pathogens on their surfaces. As such, pathogens will, inevitably, be harvested along with the duckweed crop. If harvested plants are used green as fish feed, these bacteria experience even greater dilution and faster die-off in the fish pond. The small number of surviving pathogens consumed by fish will be digested in their guts. In instances where plants are

processed and dried, desiccation will achieve even more rapid die-off. No viable human pathogens could be cultured from dried sewage-grown duckweed meal in four years of testing.

4.7 STRUVITE

The major nutrients for agriculture are nitrogen, phosphorus and potassium. Nitrogen is abundant in the atmosphere, but not directly available to most plants; whereas, energy is required to synthesize ammonium fertilizer from atmospheric nitrogen for plants. Phosphorus and potassium are obtained from ores. It has been known that phosphorus reserves are limited and those having low cadmium contents are even more limited. Cadmium contents vary from 1 to 111 mg Cd/kg DM (dry weight) (Sauerbeck and Rietz, 1980). These nutrient limitations have created more interest in nutrient recycling rather than treatment of nutrient waste streams. In particular, struvite recovery has become an emerging technology for nitrogen and phosphorus recycling. It was reported that Japanese fertilizer industries purchased struvite, obtained from sewage sludge, at a price of \$0.20/kg struvite in 2002. Conversely, calcium phosphate recovery from liquid waste is technically possible, but it might not be the most efficient way (Wilsennach *et al.,* 2003).

Conventional use of piggery wastes for fertilizing crops results in eutrophication and an accumulation of phosphorus in the soil as mentioned previously. Baker *et al.,* (1994) reported that manure applications designed to meet crop N requirements can over apply phosphorus in the soil by 2 to 3 times and cause eutrophication. Thus, an alternative to reducing waste application rates is to decrease the phosphorus concentration of waste by the precipitation of insoluble phosphate minerals. A promising method is the precipitation of magnesium ammonium phosphate ($MgNH_4PO_4.6H_2O$), commonly known as struvite or MAP. Struvite is formed as Mg^{2+}, NH_4^+, and PO_4^{3} reacts in 1:1:1 molar ratios as expressed in Equation (4.5). The solubility product (Ksp) of struvite is 7.08 x 10^{-14} moles/L as expressed in Equation (4.6).

$$Mg^{2+} + NH_4^+ + PO_4^{3-} + 6H_2O \rightarrow MgNH_4PO_4.6H_2O \quad (4.5)$$

$$Ksp = (Mg^{2+})(NH_4^+)(PO_4^{3-}) = 7.08 \times 10^{-14} \quad (4.6)$$

4.7.1 Reaction Time and pH

Swine manure slurries react fast enough to make struvite within 10 minutes at a pH of 8.5 and 40 minutes at pH 7.5. According to Burns *et al.,* (2003), a long residence time in a field scale precipitation unit was unnecessary. Since CO_2 accumulation in stored pig wastewater depresses pH, aeration is provided to increase the pH that will promote the desired struvite formation (Burns *et al.,* 2003; Heinzman and Engel, 2006). A proper Mg:PO_4 molar ratio is necessary to increase struvite formation. Mg $(OH)_2$ is very insoluble, so magnesium oxide (MgO) was used to increase struvite formation. MgO has an additional benefit of increasing the pH to aid in the struvite reaction (Beal *et al.,* 1999). However, magnesium chloride ($MgCl_2.6H_2O$) is soluble, easy to handle and requires less reaction time to bring Mg^{+2} into solution even though it depressed pH. Heinzman and Engel (2006) used $MgCl_2$ and recovered struvite from pig waste slurries using a centrifugal type grit removal system.

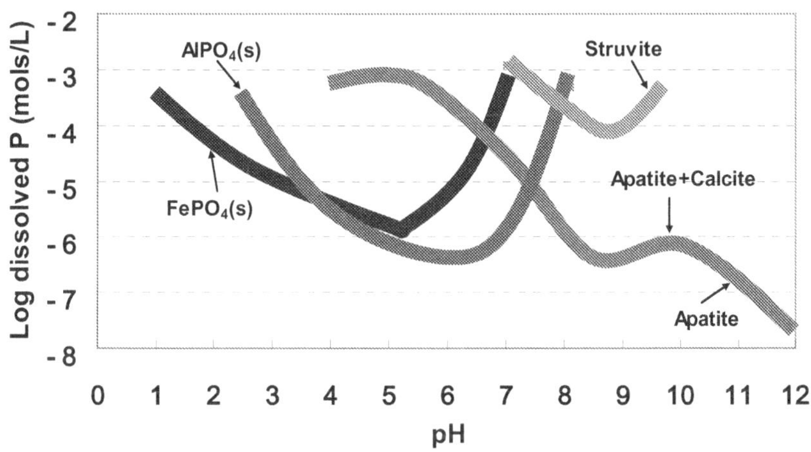

Figure 4.16 Equilibrium solubility diagrams for Fe, Al and Ca phosphates (Ferguson *et al.,* 1973) and struvite (Nelson *et al.,* 2003).

Figure 4.16 shows various phosphate precipitations occurring at different pHs. It shows the pH range required for struvite as well as those required for apatite formation with Mg and Ca. If the pH is less than 7, there will be no struvite precipitation. If pH is higher than 10, there will be $Mg_3(PO_4)_2.4H_2O$ or $Mg(OH)_2$ as shown in Figure 4.17 (Hirasawa 1996). At medium pH, there will be struvite and calcium hydroxyappatite (HAP). When Ca:Mg ratio increases,

HAP increases, and while the ratio decreases,, MAP increases (Nicolaos *et al.*, 2000).

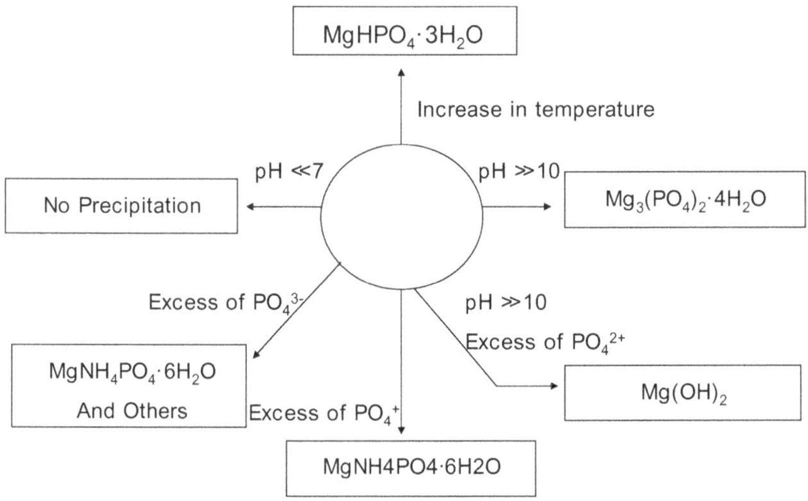

Figure 4.17 Different phosphate precipitations at different pH by Hirasawa (1996).

4.7.2 Molar Ratio

Stochiometrically, the required molar ratio of $Mg:NH_4:PO_4$ is 1:1:1 for struvite formation. Generally, Mg is limited in swine wastewater for struvite formation. The required $Mg:PO_4$ ratio can be increased to overcome the effects of complexing agents that can bind with magnesium, such as organic materials (Schuiling and Andrade, 1999). Burns *et al.,* (2003) conducted experiments to determine the required molar ratios of $Mg:PO_4$ with two different concentrations of high (590 mg/L) and low (157 mg/L) PO_4. They found 1.6:1 was sufficient to overcome the binding of organic complexion agents to the additional magnesium. Nelson *et al* (2003) obtained similar results as shown in Figure 4.18.

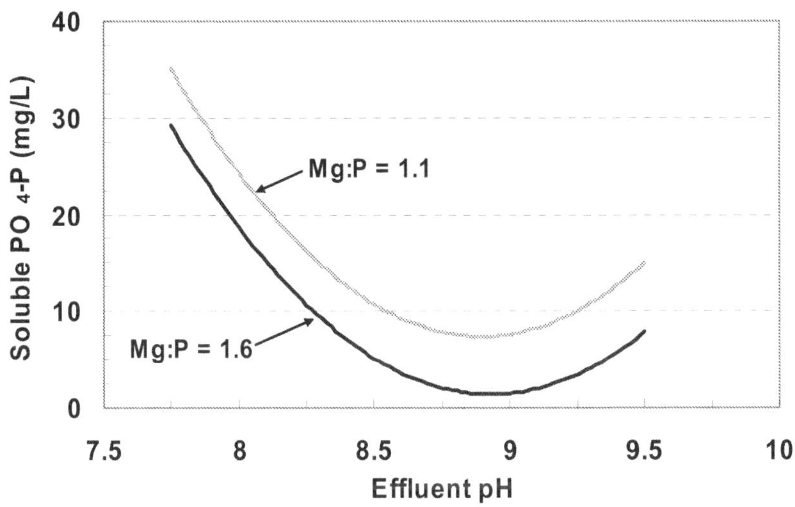

Figure 4.18 Dissolved PO_4-P concentrations from swine lagoon effluent with various Mg:P ratios after equilibrating for 24 hours (Modified from Nelson *et al.*, 2003).

4.7.3 Seeding

It was found that seeding enhanced phosphorus removal and the sizes of precipitates, but the precipitates was not still large enough to warrant enhanced recovery (Burns *et al.*, 2003). Shimamura *et al.*, (2003), however, hypothesized that the excessive growth of precipitate (crystal) reduced the reactive surface area and decreased the recovery ratio, and they suggested a two-reactor system. The main reactor facilitates particle growth from 0.5 to 1.5 mm. The next or subreactor receives fine particles in the 0.05 to 0.25 mm size range which are floating in the upper portion of the main reaction tank. When they are transferred into the subreactor, they grow up to about 0.3 mm. These grown MAP particles are recycled into the first or main reactor where they are used as the seed crystals.

Yoshino *et al.*, (2003) show that high struvite production was possible with a single continuously stirred tank reactor using a draft tube with low G value of 4 per second, and a peripheral section, where generated products are separated by gravity. The pH level was adjusted automatically by a pH controller. A paddle agitator was installed in the inner tube of the draft tube to create a descending current in the inner tube and an ascending current in the outer tube of the draft tube. Thus, the influent, Mg^{2+} solution and struvite were mixed well in the reactor.

4.7.4 Composition of Precipitates

The theoretical molar ratio of $Mg:NH_4:PO_4$ of pure struvite is 1:1:1, but the molar ratio of the precipitates from unaerated swine slurries was 1:0.74:3.2, which suggests the precipitates include HAP (Burns *et al.,* 2003). In contrast, the wasted sludge produced from a biological nutrient removal process showed that the molar ratios of $Mg:NH_4:PO_4$ were respectively 1:0.75:2.2 for 25°C and 1:0.72:2.1 for 35°C, suggesting the precipitates included slightly more HAP (Choi *et al.,* 2006).

4.8 DIRECT URINE REUSE

Direct application of urine separated from human discharge to agricultural areas has been proposed as a sustainable future use for municipalities, since urine can be a good nitrogen fertilizer containing less cadmium than in the chemical fertilizer form. Since the composition of animal urine is very similar to human urine, this sustainable concept may be applicable to animal waste management needs. Table 2.5 shows the chemical compositions of urines from human, cattle and pigs. Pig urine is more similar to that of human. Fittschen and Hahn (1998) observed how urea is decomposed during storage. As time passed, ammonium in urine is produced as shown in the following Equation:

$$CO(NH_2)_2 + 3\,H_2O \rightarrow 2\,NH_4^+ + HCO_3^- + OH^- \quad (4.7)$$

After being discharge from the human body, the pH of urine increases within about a month to the range of 9.2 to 9.5 resulting in phosphate reduction from 170 to a range of 100 to 110 mg/L, and NH_4-N increase to 3,500 mg/L (Gethke *et al.,* 2006). A similar change may be expected from pig urine. With human urine; ammonia can be recovered with struvite and ammonia stripping. The optimum pH was 9.4 to 9.7 (minimum 8.1) and Mg:P ratio was 1:5 (Harada *et al.,* 2006). After phosphorus is recovered, ammonia can be stripped into sulfuric acid at pH 1) to be recovered as $(NH_4)_2SO_4$. When the pH of sulfuric acid reaches 6, the acid will be replaced with fresh acid at pH 1. With 3,500 mg/L NH_4-N in urine, about 25 to 35 L of ammonium solution/m^3 urine could be recovered with 60°C of steam. The effluent was less than 100 mg/L NH_4-N (Gethke *et al.,* 2006).

REFERENCES

Alaerts G.J., Mahbubar R. and Kelderman P. (1996) Performance analysis of a full-scale duckweed-covered sewage lagoon, *Wat. Res.* **30**(4), 843-852.

Amy G. and Drewes J. (2005) Soil aquifer treatment (SAT) a natural and sustainable WWRR tech: Fate of Efom and trace organic compounds, *IWA Specialty Conference on Water Reclamation and Reuse for Sustainability*, Cheju, Korea.

An S (2002). *Nitrogen transformation in thermophilic aerobic digestion*, MS thesis, Department of Environmental Engineering, Korea University, Seoul, Korea.

Baguer A J, Jensen J, Krogh P H (2000) Effects of the anitibiotics oxytetracycline and tylosin on soil fauna, *Chemosphere*, 40, 751-757.

Baker J C, Zublena J P, Capbell C R (1994) *Agri-waste management: Livestock manure production and characterization in North Carolina*, North Carolina cooperative extension service, North Carolina State University

Beal L J, Burns R T and Stalder K J (1999) Effect of anaerobic digestion of struvite production for nutrient removal from swine waste prior to land application, Presented at ASAE International meeting in Toronto, Canada, paper No. 994042.

Britannica Junior Encyclopedia (1979). Hog. 171-173, Encyclopedia Britannica, Inc., Chicago, London, Toronto

Brown L, Renner M, and Halweil B (2000) *Vital signs 2000: The environmental trends that are shaping our nation*, 45, W W Norton, New York

Bull L S (2005) *Innovative sustainable systems utilizing economical solutions (ISSUES),* NC State University, Raleigh, NC

Burns R T, Moody L. B, Celen I and Buchanan J R (2003) Optimization of phosphorus precipitation from swine manure slurries to enhance recovery, *Wat. Sci. Technol.* **48**(1), 139-146.

Campagnolo E R, Johnson K R, Karpati A, Rubin C S, Kolpin D W, Meyer M T, Esteban J E, Currier R W, Smit K, Thu KM, McGeehin M (2002) Antimicrobial residues in animal waste and water resources proximal to large-scale swine and poultry feeding operations, *Sci Total Environ,* 299, 89-95.

Campos J L, Garrido J M, Mendez R, and Lema J M (2001) Effect of two broad-spectrum antibiotics on activity and stability of continuous nitrifying system, *Applied biochemistry and biotechnology*, 95, 1-10.

Chee-Sanford J C, Aminov R I, Krapac I J, Garrigues-Jeanjean N, Mackie R I (2001) Occurrence and diversity of tetracycline resistance genes in lagoons and groundwater underlying two swine production facilities, *Appl. Environ. Microbiol.* 67, 1492-1502.

Cho M K, Tran H T, Kim D H, Jia Y H and Ahn D H (2006 F). Investigation of the influence of present of antibiotics used in the efficiency of biological piggery wastewater treatment plant, *Proc of annual joint conference of Korean society of water and wastewater and Korean society of water quality*, Daegu, Korea (in Korean).

Cho S B (2001) Resource recovery from auto thermal aerobic digestion of piggery waste, Journal of Korea Organic Resource Recycling Association. Vol. 9, No.3, 27-32 (in Korean)

Choi E, Kim D, Eum Y, Yun Z and Min K S (2005) Full scale experience with nitrogen removal from piggery waste, *Wat. Environ. Res.* **77**(4), 381-388.

Choi E, Yu Y B , Cui M C , Yun Z and Min K S (2006) Effect of biologically mediated pH change on phosphorus removal in BNR system for piggery waste treatment, *Proceedings of the 7th IWA Agro 2006*, 191, Amsterdam

Chung C H (1981) *Effect of nightsoil storage in army compounds*, MS thesis, Department of civil and environmental engineering, Korea University, Seoul, Korea

Conservation Practice Standard (2003) *Draft for anaerobic digester-controlled and ambient temperatures,* code 365 and 366, US Natural resources conservation service.

De Liguoro M, Cibin V, Capolongo F, Halling-Sorensen B, Montesissa C (2003) Use of oxytetracycline and tylosin in intensive calf farming: Evaluation of transfer to manure and soil, *Chemosphere*, 52, 203-212.

Eby H J (1964) "Anaerobic lagoons"-theory and practice. *2nd National Symposium on Poultry Waste Management Proc.*, 77-91.

Eum Y J (2006) *Modeling of nitrogen and phosphorus removal from piggery waste*, PhD thesis, Department of Civil and Environmental Engineering, Korea University, Seoul, Korea.

Ferguson J F, Eastman J and Jenkins D (1973) Calcium phosphate precipitation at slightly alkaline pH values, *JWPCF*, 45, 620-627.

Fittschen I and Hahn H H (1998) Characterization of the municipal wastewater part human urine and a preliminary comparison with liquid cattle excretion, *Wat. Sci. Technol.* **38**(6), 9-16.

Gethke K, Herbst H, Montag D, Bruszies D, Pinnekamp J (2006) Phosphorus recovery from human urine, presented at *IWA WWC*, Beijing

Golet E M, Xifra I, Siegrist H, Alder A C and Giger W (2003) Environmental exposure assessment of flouroquinolone antibacterial agents from sewage to soil, *Env Sci and Tech*, 37, 3243-3249.

Haelzle J J (2005) Concentrated animal feeding operations (CAFOs) and manure management – An alternative approach, *Animal agriculture and processing: Managing environmental impact*, Air and waste management association, WEF, St. Louis, MO.

Halling-Sorensen B (2000) Algal toxicity of antibacterial agents used in intensive farming, *Chemosphere,* 40, 731-739.

Hamscher G, Sczesny S, Hoper H and Nau H (2002) Determination of persistent tetracycline residues in soil fertilized with liquid manure by high-performance liquid chromatography with electrospray ionization tandem mass spectrometry, *Analytical chemistry*, 74, 1509-1518.

Harada H, Shimizy Y, Miyagoshi Y, Matsui S, Matsuda T and Nagasaka T (2006) Predicting struvite formation for phosphorus recovery from human urine using an equilibrium model, presented at *IWA WWC*, Beijing

Haustein A, Gilman R, Skillicorn P (1987) *The safety and efficacy of sewage-grown duckweed as feed for layers, broilers and chicks*, report to USAID science advisor.

Heinzmann B and Engel G (2006) Induced magnesium ammonia phosphate precipitation to prevent in crustations and measures for phosphorus recovery, presented at *IWA WWC* in Beijing

Hirasawa I (1996) Study on the recovery of ions in wastewater by crystallization, *Memoirs of the School of Science and Engineering,* Waseda University, No.60

Huang C H, Renew J E, Smeby K L, Pinkston K and Sedlak D L (2001) Assessment of potential antibiotic contaminants in water and preliminary occurrence analysis, *Proc of the 2nd Int conference on pharmaceuticals and endocrine disrupting chemicals in water*, October 9-11, Minneapolis, MN

Humenik F (2005) Introduction in *Animal waste containment in lagoons* edited by Reddi, L.N., 6, ASCE

Humenik F J and Overcash M R (1976) Design criteria for swine waste treatment systems, Rep No. EPA – 600/2-76-233 EPA, Ada.

Hwang I S and Min K-S (2006) Piggery waste treatment using improved MLE process in full scale, *J KSWQ*, **22**(5), 895-904.

Jewell W J Kabrick R M (1980) Autoheated aerobic thermophilic digestion with aeration, *Wat. pollut. control fed,* 52, 512-523.

Kim Y C, Choi E (1991) Autothermal thermophilic aerobic digestion of nightsoil, *J. KSWPRC August 170-175*

Kolpin D W, Furlong E T, Meyer M T, Thruman E M, Zaugg, S D, Barber L B and Buxton H T (2002) Pharmaceuticals, hormones, and other organic wastewater contaminants in US streams, 1999-2000: A national reconnaissance, *ES&T*, 36, 1202-1211.

Kolz A C, Moorman T B, Ong S K, Scoggin K D and Douglass E A (2005) Degradation and metabolite production of tylosin in anaerobic and aerobic swine-manure lagoons, *Wat. Env. Res.* **77**(1), 49-57.

Korean RDA (2002) *Guidelines for the use of liquid fertilizer.* Technology transfer 11-1390093-000082-01. www.rda.go.kr

Kvarnstrom E, Schonning C, Carisson-Reich M, Gusttafsson M and Enocksson E (2003) Recycling of wastewater-derived phosphorus in Swedish agriculture, *Wat. Sci. Technol.* **48**(1), 19-25.

Lampert C (2003). Selected requirements on a sustainable nutrient management, *Wat. Sci. Technol.* **48**(1), 147-154.

Lapara T and Alleman J E (1999) Thermophilic aerobic biological wastewater treatment, *Wat. Res.* **33**(4), 895-902.

Lee J H (2006). *Piggery waste treatment with ABR and BNR processes*, MS thesis, Department of Civil and Environmental Engineering, Korea University, Seoul, Korea.

Lue-Hing C, Zenz D R and Kuchenrither R (1992) *Municipal sewage sludge management*, 273, Technomic.

Natural Resources Conservation Service of North Carolina (1989) *Waste treatment lagoons;* Standard 359, Raleigh, NC.

Nelson N O, Mikkelsen R L and Hesterberg D L (2003) Struvite precipitation in anaerobic swine lagoon liquid: effect of pH and Mg:P ratio and determination of rate constant, *Biresource technology,* 89, 229-236

Nicolaos C, Bouropoulos P, Koutsoukos G (2000). Spontaneous precipitation of struvite from aqueous solutions. *J. Crystal Growth*, 213, 381-388.

Official journal of the European community (1998) also from Brinton W F (2001) An international look at compost standards, *Biocycle*, 74, April

Olutayo O A (2005) Agricultural statistics and environmental issues, International water management institute-Ghana, personal communication

Park S W (2001) Upgrading composting plant: A case study, *J KOWREC*, **9**(3), 40 in Korean.

Roos K (2003) Comparative environmental benefits of anaerobic digesters, *Proceeding Anaerobic digester technology application in animal agriculture- A national summit*, Raleigh, NC, WEF.

Sauerbeck D and Rietzelastung E (1980). Zur cadmiumbelastung von minerldeungern in abhangigkeit von rohstoff und herstellungsverfaren, Landw, forschung, sdhft, 37, 685-692.

Schuiling R D and Andrade A (1999) Recovery of struvite from calf manure, *J of Environment Technology,* 20, 765

SCS (Soil Conservation Service) (1970) "Interim specifications – Disposal lagoons" Advisory ENG-S, US Dept of Agriculture, Washington DC.

Shimamura K, Tanaka T, Miura Y and Ishikawa H (2003) Development of a high efficient phosphorus recovery method using a fluidized-bed crystallized phosphorus removal system, *Wat. Sci. Technol.* **48**(1), 163.

Shuval H I (1991) Prospective epidemiological study on enteric disease transmission associated with sprinkler irrigation with wastewater: On overview in implementing water reuse, *Proceedings of the water reuse symposium IV*, Denver, CO. AWWA Foundation.

Skillicorn P, Spira W. And Journey W (1993) *Duckweed Aquaculture: a new aquatic farming system for developing countries*, The World Bank, Washington, D.C.

Skjelhaugen O J (1991). Liquid composting unit. *Proceedings CIGR seminar in Environmental challenges and solutions in agricultural engineering,* ISBN 82-7636-009, 52-57

Skjelhaugen O J and Donantoni L (1998). Combined aerobic and electrolytic treatment of cattle slurry. *Journal of agricultural engineering research*. 70(2). 209-219

Skjelhaugen O J and Sæther T (1994). *Prosjektet local økologisk avløpsløning i Etnedal kommune, erfaringer ogresultater etter første års drift (The local ecological waste water project, experiences and results from the first year)* Print no 54-94, Dept. of Agric. Engineering, Agricultural University of Norway

Skjelhaugen O J, Sæther T, Hanssen J F, Jenssen P D (1992). *Samkompostering av septikslam og husdyrgjødsel (Co-composting of sewage sludge and livestock slurry)* Print no 59-92, Dept. of Agric. Engineering, Agricultural University of Norway

Spinosa L, Leschber R, Wichmann K (2006) Sewage sludge co-management: Developments in EU regulations and characterization procedures, *Proceedings of IWA specialized conference-Sustainable sludge management: State of the art, challenges and perspectives* 11, Moscow

Sulabh International (2005) Personal communication while visiting India.

Surucu G A, Chian E S K, Engelbrecht R S (1976). Aerobic thermophilic treatment of high strength wastewaters, *Wat. Pollut. Control Fed.* 48, 669-679.

Teeter J S, Meyerhoff R D (2003) Aerobic degradation of tylosin in cattle, chicken and swine excreta, *Environ Res*, 93, 45-51

Tilati S N, Stenstrom M K(1990). Aeration-basin heat loss, *Environ. Eng.* 116, 70-86.

Tolls J (2001) Sorption of veterinary pharmaceuticals in soil, a review, *ES&T*, 35(17), 3397-3406.

USDA (2002) *Preventive practices in swine: Administration of iron and antibiotics*, US department of agriculture, Fort Collins, CO.

Wilsenach J A, Maurer M, Larsen T A and van Loosdrecht MCM (2003) From waste treatment to integrated management, *Wat. Sci. Technol.* **48**(1), 1-9.

Won S G and Ra C S (2001). The impacts of toxic chemicals in animal wastewater on biological treatment, *J of Korean society of environmental engineers*, **23**(11), 1843-1850.

Yoshino M, Yao M, Tsuno H and Somiya I (2003) Removal and recovery of phosphate and ammonium as struvite from supernatant in anaerobic digestion, *Wat. Sci. Technol.* **48**(1), 171.

Yi Y S, Kim S, An S, Choi E, Yun Z (2003), Gas analysis reveals novel aerobic deammonification in thermophilic aerobic digestion, *Wat. Sci. Technol.* **47**(10), 131-138.

5

Nutrient removal

5.1 INTRODUCTION

Animal production for human consumption is important to the global nitrogen economy because of the waste streams created and because of the demand for plant crops to produce the animals. In the United States, almost five times more crop nitrogen is fed to animals than is consumed directly by humans with plant food. A change in the US per capita meat consumption to a typical "Mediterranean diet" could result in a drop in the US nitrogen fertilizer requirements by 50%, This would result in substantial benefits to the environment because of the two-fold reduction in fertilizer use and a seven-fold decrease in animal wastes. Unfortunately, the US trend for per capita meat production (including poultry) is expected to rise from about 49 g/capita/d in 1961 to about 72 g/capita/day in 2010 (Howarth, 2004). Elsewhere in the world, per capita consumptions is steady or falling in developed countries as a whole, but rising in the developing world. The total amount of meat production in the world is rising rapidly as shown in Table 4.1. Note that globally one third of the meat production is based on grazing systems rather than crop or foraged-fed systems, but the trend is strongly toward the crop-fed system. In particular, pigs are almost raised on the crop-fed system in large scale enclosed area.

As shown on Table 2.6, a pig daily produces about 30 g N and 9 g P with about 510 g COD and a total flow of 5 L (3 L of urine and 2 L of feces). After separation of the solids, the liquid represents about 20 g N, 1.7 g P and 94 g COD/head/day (see Figure 2.5) resulting in 6,000 mg/L N, 570 mg/L P and 30,000 mg/L COD. The concentrations will vary with ambient temperature (actually the room temperature for the pig), solid-liquid separation methods and cleaning frequency of the pig houses. As the ambient temperature increases, pigs drink more water, and the concentrations will be lower. When anaerobic digestion is used to treat the mixed waste (urine and feces), COD and phosphorus concentrations are reduced greatly and only a small reduction in nitrogen concentrations is expected. Phosphorus removal in the digester is attributed to chemical precipitation with struvite and apatite. This will be discussed later in this chapter.

5.2 NITROGEN REMOVAL PROCESSES

Figure 5.1 presents various nitrogen removal processes which may be applicable to piggery waste treatment plants. The MLE (modified Luzack-Ettinger), Bardenpho process, oxidation ditch and SBR (sequencing batch reactor) processes are frequently used.

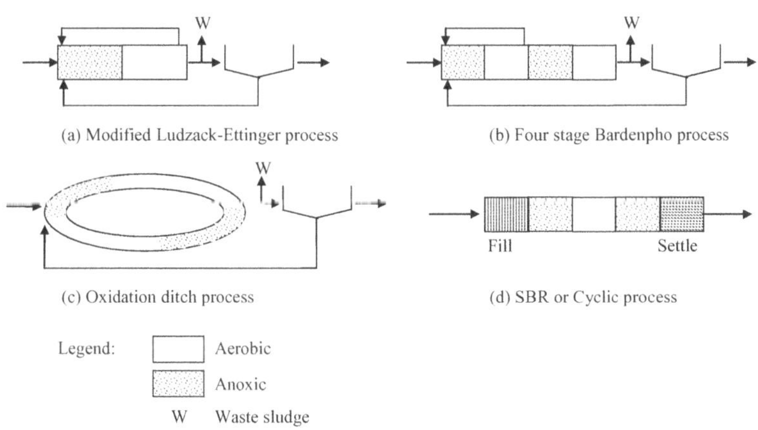

Figure 5.1 Nitrogen removal processes applicable for piggery wastewater.

5.2.1 Nitrification

Nitrogen removal is accomplished using the biological nitrification and denitrification processes. Nitrification is a two stage process involving ammonium oxidizing bacteria (AOB) and nitrite oxidizing bacteria (NOB);

$$NH_4\text{-}N + 1.5\ O_2 \rightarrow NO_2\text{-}N + H_2O + 2H^+ \tag{5.1}$$

$$NO_2\text{-}N + 0.5\ O_2 \rightarrow NO_3\text{-}N \tag{5.2}$$

These equations show that in order to nitrify 1 g of NH_4-N to NO_3-N, 4.6 g O_2 is required, while 3.4 g O_2 will be required for nitrite nitrification. This results in 25% of savings in O_2 requirement with NO_2-N nitrification. In terms of alkalinity, nitrite nitrification requires 7.1 g Alk./g NH_4-N, while practically no alkalinity is required for NO_2-N to NO_3-N conversion. This fact suggests that sufficient alkalinity must be presented to achieve stable nitrite nitrification. If the alkalinity were exhausted during nitrite nitrification, NO_2-N conversion from NH_4-N would be stopped. However, the conversion to NO_3-N conversion can still proceed if NO_2-N is available.

During the denitrification process where sufficient carbon is available, alkalinity is recovered at 3.6 g Alk./g NO_2-N or g NO_3-N removed. As far as oxygen is concerned, 2.86 g oxygen is recovered per g NO_3-N denitrification, since 8g of oxygen (1/4 x 32 g of O_2/mole) is equivalent to 2.8 g NO_3N, and 1.71g oxygen is recovered from NO_2-N denitrification with the same computation.

$$\frac{1}{4}O_2 + H^+ + e^- \rightarrow \frac{1}{2}H_2O$$

$$\frac{1}{5}NO_3^- + \frac{6}{5}H^+ + e^- \rightarrow \frac{1}{10}N_2 + \frac{3}{5}H_2O \tag{5.3}$$

$$\frac{1}{3}NO_2^- + \frac{4}{3}H^+ + e^- \rightarrow \frac{2}{3}H_2O + \frac{1}{6}N_2 \tag{5.4}$$

There are many factors associated with nitrification. These include: temperature, dissolved oxygen, pH, inhibiting substances, influent COD/N ratios and SRT (solid retention time).

(1) Temperature

The nitrification process is strongly dependent on temperature (USEPA, 1993). Nitrification occurs over a range of approximately 4 to 45°C, with an optimum at about 35°C for *nitrosomonas* with a 32 to 42°C optimum range for *nitrobacter*. Figure 5.2 shows how AOB and NOB competes at different temperatures (Beline *et al.,* 2007). When piggery waste is applied to an aerobic system at a higher load, the reactor temperature will be increased particularly during the summer due to the heat released by the aerobic decomposition of organic matter as discussed in Chapter 4 (composting or thermophilic aerobic digestion). The critical lower temperature affecting growth rate for *nitrifiers* is in the range of 12 to 15°C . For design purposes, an acceptable Arrhenius-type expression for the effect of temperature on the maximum growth rate of nitrifiers over a range of 5-30°C for domestic sewage is (US EPA, 1993):

$$Maximum \quad U_N = 0.47 \, e^{0.097(T-15)} \tag{5.5}$$

where: T = temperature,°C

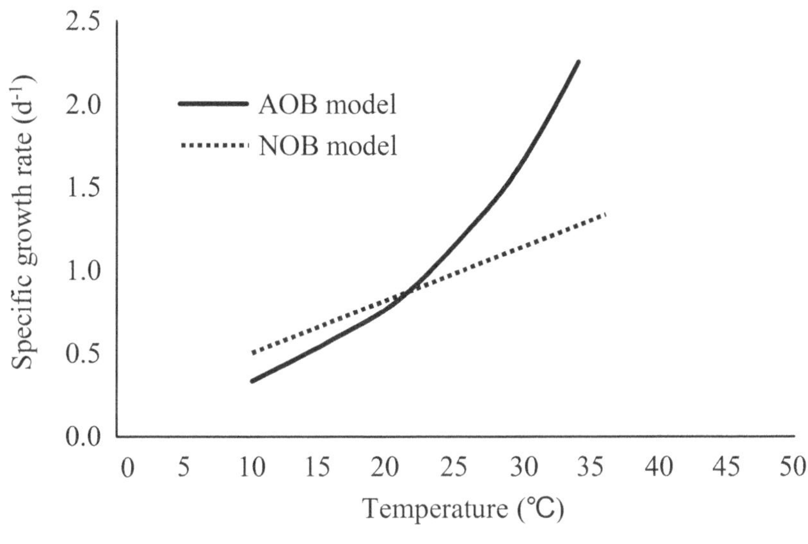

Figure 5.2 Effect of temperature on the specific growth rates of ammonia oxidizing bacteria (AOB) and nitrite oxidizing bacteria (NOB). The growth rate of AOB exceeds that of NOB at higher temperatures (Modified from Beline *et al.*, 2006).

(2) Dissolved Oxygen

The concentration of DO has a significant effect on the rates of nitrifier growth and nitrification in biological treatment systems. The nitrogen manual (USEPA, 1993) indicates that the intrinsic growth rate of AOB is not limited at DO levels above 1.0 mg/L, but that DO levels greater than 2.0 mg/L may be required in practice. When designing the aeration or oxygen addition component of a suspended growth nitrification system, it is recommended that a minimum DO level of 2.0 mg/L be specified at all times throughout the biological reactor to prevent ammonia bleed-through at peak ammonia loadings.

(3) pH and Alkalinity

Raw piggery wastes generally have a pH higher than 7 to 9. However, untreated piggery wastes can often show a lower pH, but the pH will be increased after mixing and short aeration resulting in the removal of CO_2 from the waste (see Figure 5.3) and the hydrolysis of urea to NH_4OH (see Equation 4.7). Figure 5.4 shows that pH has a significant effect on the rate of nitrification. Since the pH range in nitrification and denitrification systems for piggery waste is typically 7 to 8 (except in some cases which are induced from diluted influent concentrations with high nitrogen and low COD, namely a low COD/N ratio), the nitrification is easily obtained without an artificial pH increase.

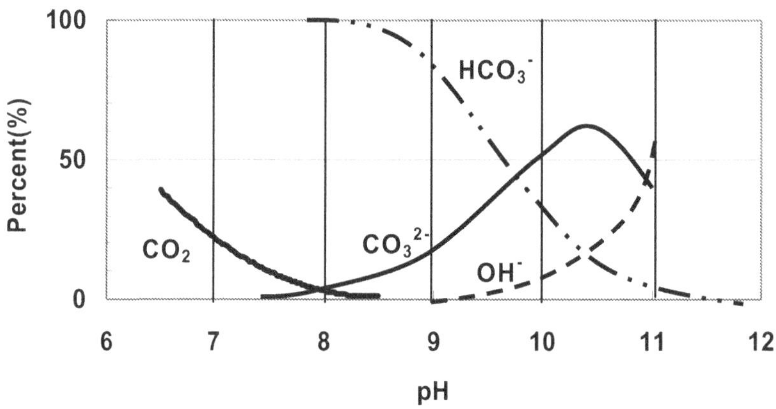

Figure 5.3 Relationship between carbon dioxide and forms of alkalinity at various pH (US EPA, 1993).

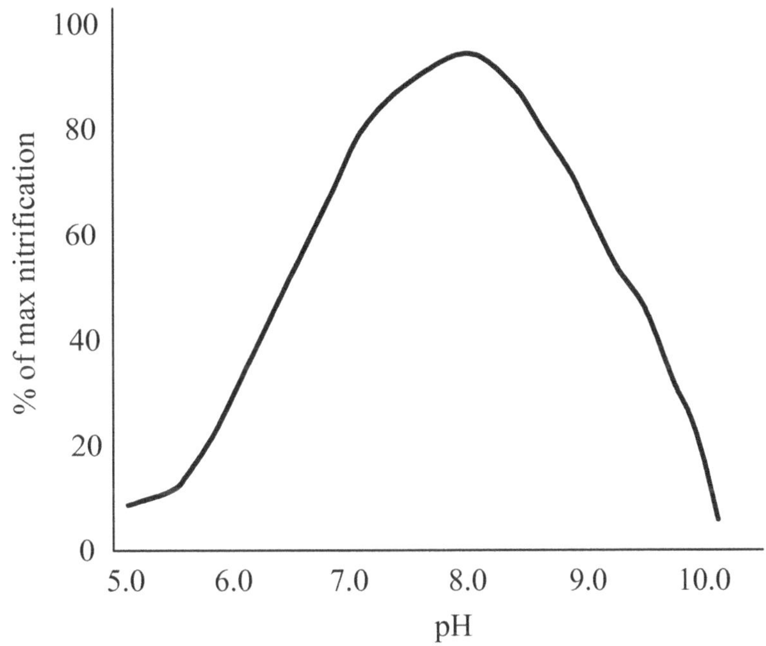

Figure 5.4 Effect of reactor pH on nitrification rates (Modified from US EPA, 1993).

There can be a difference between actual and measured alkalinity in piggery waste because volatile fatty acids (VFA) present in piggery waste depress the actual alkalinity. The actual alkalinity can be computed from the measured bicarbonate alkalinity and VFA COD (as acetate) using the equation given by Snoeyink and Jenkins (1980):

$$\text{Total alkalinity} = \text{bicarbonate alkalinity} + 0.083\ (\text{VFA COD}/1.067) \qquad (5.6)$$

Where, 0.833 = alkalinity equivalent of acetic acid, and
1.067 = COD equivalent of acetic acid (CH_3COOH)

With the actual alkalinity, a mass balance can be made along with nitrification and denitrification reactions. Figure 5.5 compares the measured and theoretical alkalinity requirements along with influent TCOD/TKN ratios. Nitrification requires 7.1 g Alk/g NH_4-N removed. However, Figure 5.5 shows the alkalinity requirement decreases as the influent COD/TKN ratios increases.

The minimum alkalinity/TKN requirement is as small as 3.6 at an influent COD/TKN ratio of about 6.

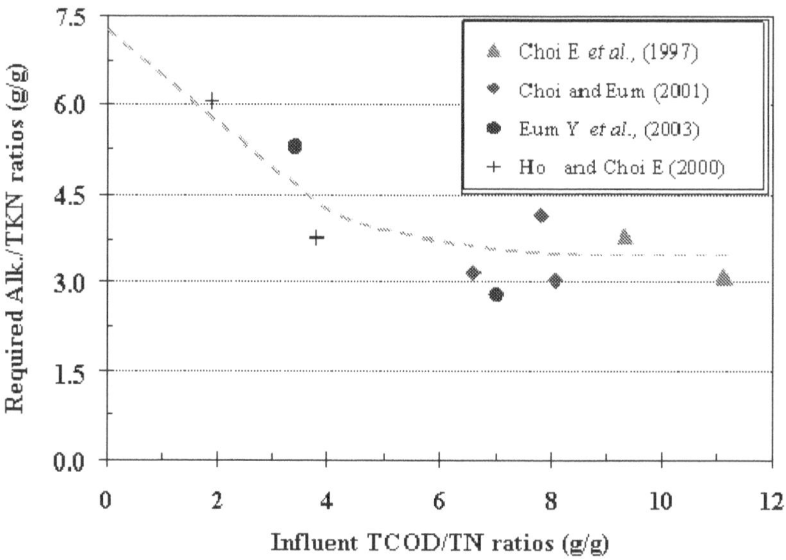

Figure 5.5 Alkalinity requirements for nitrification and alkalinity recovered from denitrification for piggery waste. Alkalinity requirement decreases due to the recovered alkalinity from denitrification, which increases as influent COD/TKN ratios increase.

(4) Inhibiting substances

Inhibiting substances such as heavy metals and organic chemicals are considered to be absent in piggery wastes. As discussed in Chapter 4, antibiotics, promoters and other medicinal substances are not considered to inhibit nitrification to any great extent. However, it has been known that nitrifiers are sensitive to certain forms of nitrogen. Unionized ammonia (or NH_3, free ammonia and FA), and unionized nitrous acid (or HNO_2, free nitrous acid and FNA) are believed to be inhibitory to nitrifiers above certain concentrations. Anthonisen *et al.,* (1976) reported that FA begins to inhibit *Nitrosomonas* at a concentration of 10 to 150 mg/L and *Nitrobacter* in the range of 0.1 to 1.0 mg/L. FNA begins to inhibit *Nitrosomonas* and *Nitrobacter* at concentrations of 0.2 to 2.8 mg/L. The FA and FNA concentrations are directly correlated to pH and temperature, and the concentrations of ammonia plus ammonium and nitrite plus nitrous acid, respectively. FA and FNA are present in accordance with the following equilibrium reactions:

$$NH_4^+ + OH^- \leftrightarrow NH_3 + H_2O$$
$$H^+ + NO_2^- \leftrightarrow HNO_2$$

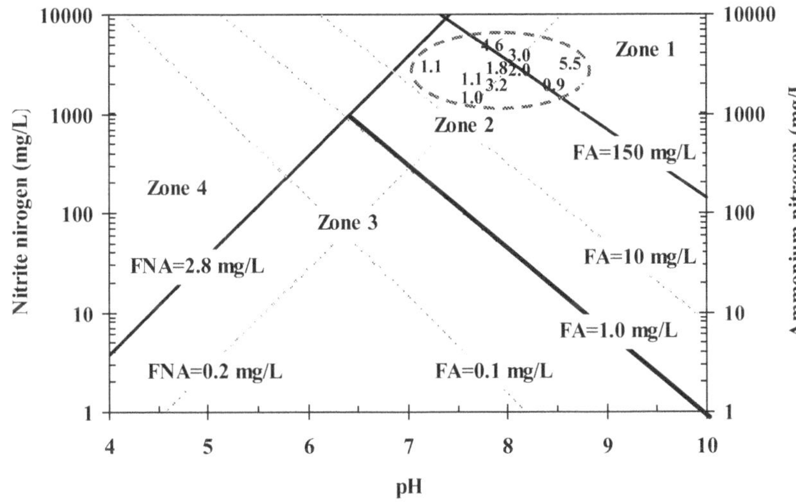

Figure 5.6 Inhibition of *Nitrobacter* and *Nitrosomonas* by free ammonia (FA) and free nitrous acid (FNA) with data points taken from a nitrification study of piggery wastes (Modified from Anthonisian *et al.,* 1976). The circled data points with values identify influent nitrogen concentrations in g/L as shown in Figure 5.8.

- Zone 1: Inhibition of Nitrobacter and Nitrosomonas by FA
- Zone 2: Nitrobacter inhibition ⇒ nitrite nitrification by FA
- Zone 3: Complete nitrification
- Zone 4: Nitrobacter inhibition ⇒ nitrite nitrification by FNA

Accumulation of nitrite is beneficial for nitrogen removal from piggery waste. Considering inhibition effects cited by Anthonisen *et al*, (1976), Choi *et al*, (2005) proposed a simple relation between nitrite nitrification and nitrate nitrification with the final pH at the end of oxic stage in SBR (designated as oxic end pH in some place in this book) and nitrogen loads as shown in Figure 5.7. It shows that nitrite nitrification occurs at higher pHs with lower nitrogen loads or at lower pHs with higher nitrogen loads. The proposed nitrification rates for the design are 33 to 36 mg/L/hr for nitrite nitrification and 13 mg/L/hr for nitrate nitrification (Choi *et al.,* 2004). However, there are other factors which affect nitrite and nitrate nitrifications; e.g., SRT and DO (besides pH and nitrogen loads) (Jenicek *et al.,* 2004).

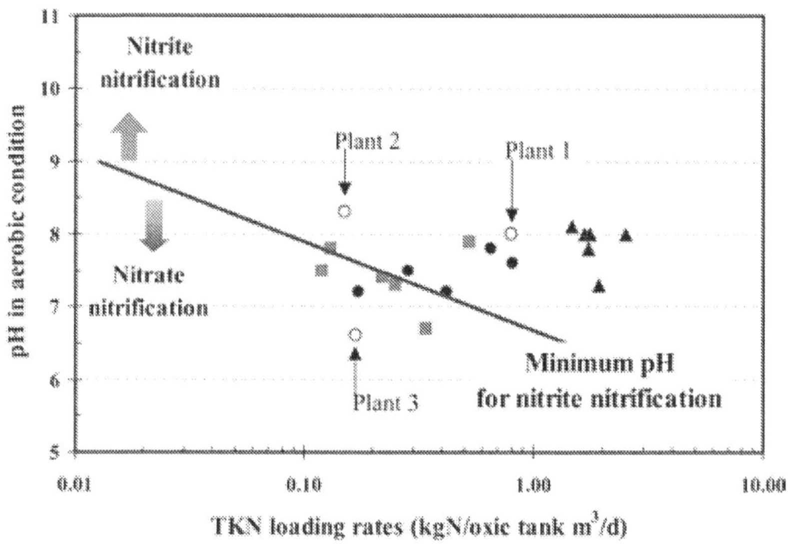

Figure 5.7 Nitrite nitrification with TKN loading and oxic pH (Choi *et al.,* 2005). Reference data as follows: ■ Eum and Choi (2002); ▲ Hwang *et al.,* (2000); ● Gil *et al.,* (2002). (Plants 1, 2 and 3 are from Chapter 7.4.4).

(5) COD/N ratios

The influent biodegradable COD/nitrogen ratio is one of the critical factors affecting the design of nitrification systems. Normally, organics in piggery waste are mostly biodegradable, but that depends on the influent concentrations. About 10% of the influent COD and 1% of the influent nitrogen are considered to be nonbiodegradable fractions as shown in Figure 3.6 using the ASM model (2000). With high COD/N ratios, heterotrophic bacteria growth predominates in the reactor and the nitrifiers have to compete with them for oxygen. This is particularly true when attempting to control the MLSS at a desired level. Proper consideration for the nitrifiers must be taken to insure that are not washed out at a given SRT.

EU communities use COD and the US uses BOD for the design of biological treatment systems. Korea and Japan use BOD and COD_{Mn}. COD_{Mn} results are obtained using $KMnO_4$ instead of $K_2Cr_2O_7$. The IWA design model for activated sludge (ASM Model, 2000) includes fractionated COD concentrations. If fractionated COD values are not available, the following COD and BOD relationships may be utilized for design purposes assuming that the influent COD to the biological process is 20,000 mg/L COD during winter after proper solids removal:

Total biodegradable COD = about 90% of total COD of 20,000 mg/L= 18,000 mg/L
Ss = VFA COD = soluble BOD = about 28% of total COD = 20,000 × 0.28 = 5,600 mg/L
Total BOD × 1.72 = total biodegradable COD
→ Total BOD = 10,400 mg/L
Particulate BOD = 10,400 – 5,600 = 4,800 mg/L

(6) SRT

The Monod type Equation given below represents the actual required SRT for a given system (US EPA, 1993):

$$U_N = \frac{1}{SRT} = 0.47\, e^{0.098(T-15)} \times \left[\frac{N}{(K_N + N)} \times \frac{DO}{(K_O + DO)} \times [1 - 0.833(7.26 - pH)] \right]$$

(5.7)

where:
μ_N: Specific growth rate of *Nitrifier*, /d
N: NH_4-N concentration, mg/L
K_N: NH_4-N half saturation coefficient for (AOB), mg/L
DO: 3 mg/L at 25°C
K_o: Half saturation coefficient for oxygen, mgNH_4-N/L
T: Temperature, °C

The NH_4-N levels used with this equation for different pH conditions (7.2 and 7.8), K_N (1.34 to 7.08) and K_o values (0.3 to 1) assume that ammonium oxidizing bacteria (AOB) are dominate (Eum and Choi, 2002; Beline *et al.,* 2006). The resultant minimum SRT range for nitrification was calculated to be 0.9 to 1.7 d. However, the predicted values are always lower than the measured, since the actual operating SRT is always greater than the minimum SRT. Actually, the influent piggery waste concentration, which varies greatly, affects the effluent NH_4-N levels as shown in Figure 5.8. The influent values are TKN concentrations. At 10 day oxic HRT, the effluent NH_4-N levels are about 20 mg/L with a 1 g TKN/L influent, about 35 mg/L with a 2 g TKN/L influent and 115 mg/L with a 6 g N/L influent. These data are from a completely mixed oxic reactor where the SRTs are about four times the HRTs (Yu, 2007). Actually, multicompartment oxic reactors may be used for strong nitrogenous waste and the first order kinetic model can be used for this design approach. The predicted effluent NH_4-N concentrations for different influent TKN concentrations are shown in Figure 5.8 using a nitrification rate (K_{mn}) of 5/d at 25°C. Table 5.1 presents proposed design equations for piggery wastes that can be used to predict effluent NH_4-N concentrations with given influent concentrations.

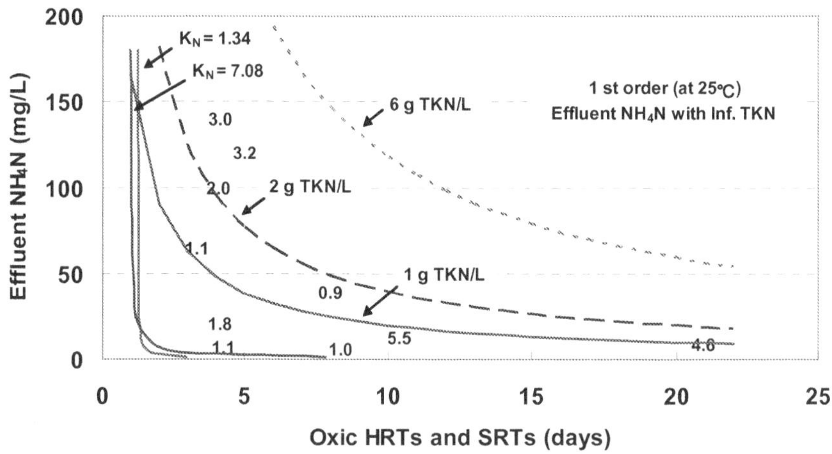

Figure 5.8 Predicted effluent NH_4-N concentrations using a first order kinetic model.

Table 5.1 Proposed design equations for nitrification of piggery wastewater.

$$NH_4\text{-}N_{eff} = \frac{N_i}{K_{mn} \times HRT_{ox} + 1} \quad (5.8)$$

$$HRT_{ox} = \frac{(N_i / NH_4 - N_{eff}) - 1}{K_{mn}} \quad (5.9)$$

where

NH_4-N_{eff} = Computed effluent NH_4-N (mg/L)
N_i = Influent TKN (mg/L)
K_{mn} = Nitrification rate (/d)
HRT_{OX} = Oxic HRT

$$K_{mn} = K_{25} \times (1.047)^{T-25} \quad (15 \leq T \leq 40°C) \quad (5.10)$$

K_{25} = 5/d @ 25°C

5.2.2 Denitrification

Unlike nitrification, a relatively broad range of bacteria can accomplish denitrification. Denitrifiers are ubiquitous in most natural environments, including piggery wastewaters and sludges. Many of the microorganisms in activated sludge systems are denitrifiers, even in systems that are not specifically designed for denitrifying. These organisms are facultative and they can use oxygen, nitrite or nitrate as their terminal electron acceptor. Denitrifiers are heterotrophic microorganisms which can proliferate in aerobic systems because of their ability to use oxygen and efficiently oxidize organic matter. The denitrification process involves transferring electrons from a reduced electron donor (e.g. an organic substrate) to an oxidized electron acceptor (e.g. nitrate or nitrite). Nitrate or nitrite serves as a substitute for oxygen in this chain with only small modification to the metabolic system of the bacteria (US EPA, 1993). However, denitrifiers are differentiated as the organisms (X_{HD}) utilizing NO_2-N or NO_3-N in the absence of DO, which is called an anoxic environment. The rate of denitrification is dependent on temperature, DO, organic concentration or COD/NOx-N ratio, pH and the available nitrified forms. Figure 5.9 shows how temperature affects COD and NH_4-N removal in an aerobic system. It shows how these heterotrophic microorganisms are less sensitive than nitrifiers.

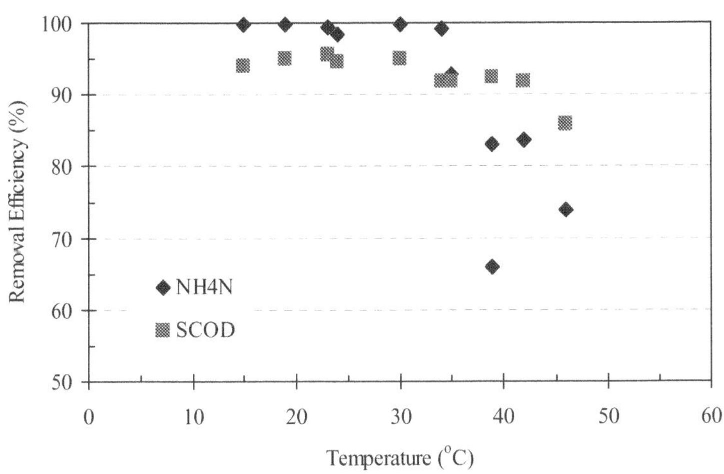

Figure 5.9 SCOD and NH_4-N removal efficiencies with temperature (modified from Choi *et al.*, 2003). SCOD denotes soluble COD.

Table 5.2 Design equations for denitrification (modified from US EPA, 1993; IWA ASM model, 2000).

(1) Denitrification using the Monod equation

$$q_D = \hat{q}_D \left(\frac{S}{Ks \times S}\right)\left(\frac{NOxN}{K_D + NOxN}\right)\left(\frac{So}{Ko + So}\right) \quad (5.12)$$

where

q_D = NO_x-N removal rate, g NO_x-N/gVSS/d

$\hat{q}_D$ = Maximum NO_x-N removal rate

S = Concentration of organic substrate, mg/L

K_s = Half saturation constant, mg/L

K_D = Half saturation constant for NO_x-N, mg/L

S_O = DO concentration, mg/L

K_O = Half saturation constant for oxygen, mg/L

(2) Carbon requirement for denitrification

$$\frac{COD}{NOx - N} = \frac{2.86}{1 - 1.134Y_{net}} \quad (5.13)$$

where

Y_{net} = Biomass net yield based on COD, $gVSS/gCOD_{removed}$

2.86 = Denitrification from NO_3-N and 1.71= Denitrification from NO_2-N

NOx-N = Nitrogen to be removed by denitrification

$$Y_{net} = \frac{Y}{1 + K_d \ SRT} \quad (5.14)$$

Y = Synthesis yield coefficient, gVSS/gCOD

K_d = Endogenous decay coefficient, /d

(3) Effluent NO_x-N concentrations when carbon is not limited

$$NO_x\text{-N effluent} = \frac{Influent \ \ NOxN}{(RAS + NRCY + 1)} \quad (5.15)$$

where

RAS = Return activate sludge ratio, Qr/Q

NRCY = NO_x-N recycle ratio

Design equations for denitrification are shown in Table 5.2. The carbon requirements for nitrogen removal can be estimated using the method proposed by McCarty *et al* (1969):

$$M = 2.47\ (NO_3\text{-}N) + 1.53\ (NO_2\text{-}N) + 0.87\ DO \qquad (5.11)$$

where:

M = methanol required, mg/L (multiply 1.5 to obtain COD)
NO_3-N = nitrate nitrogen removed, mg/L
NO_2-N = nitrite nitrogen removed, mg/L
DO = DO removed, mg/L

However, it should be noted that the estimated COD requirements based on McCarty's equation are almost 20% higher that the measured (Ho and Choi, 2000). The measured COD required/NO_2-N removed in a SBR operation with a 25 day SRT was 1.9 to 2.9 and the NO_x-N uptake rate (NUR) was 11.2 to 16.3 mg NO_2-N/L/hr (Eum and Choi, 2002).

(1) Organic Concentrations or COD/N Ratio

Sufficient carbon is required for NO_x-N removal from denitrification. Figure 5.10(a) suggests a COD/N ratio greater than six gives more than 90% nitrogen removal efficiencies. Proper oxic and anoxic volumes are required for nitrification and denitrification, respectively. Also, the volume requirements vary with COD/N ratios as shown in Figure 5.10(b).

(2) pH and Alkalinity

Figure 5.5 shows the amount of alkalinity recovered from denitrification. As indicated earlier, nitrification suppresses the pH and denitrification recovers alkalinity or increases the pH. A higher organic load stimulates denitrification; however, the increased pH demonstrated that denitrification inhibition occurred around pH 9 (Ho and Choi, 2000). Figure 5.11 shows effect of pH on denitrification rates (US EPA, 1993). The minimum COD/TN and alkalinity/TN ratios required for higher nitrogen removal are about 6 and 3.6, from Figures 5.10 and Figure 5.5, respectively.

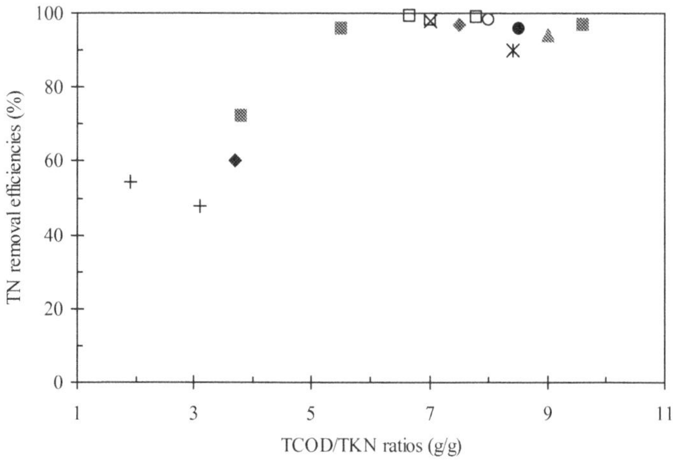

(a) TN removal efficiencies

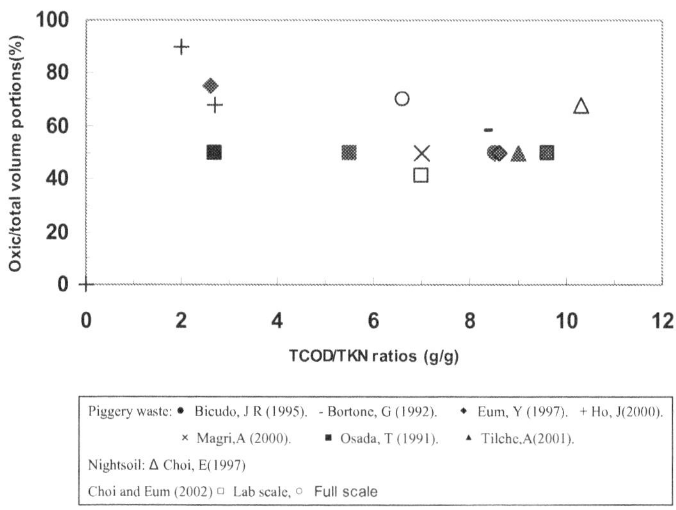

(b) Oxic/total volume portions

Figure 5.10 TN removal efficiencies and oxic/total volume portions as a function of influent TCOD/TKN ratios.

(2) pH and Alkalinity

Figure 5.5 shows the amount of alkalinity recovered from denitrification. As indicated earlier, nitrification suppresses the pH and denitrification recovers alkalinity or increases the pH. A higher organic load stimulates denitrification; however, the increased pH demonstrated that denitrification inhibition occurred around pH 9 (Ho and Choi, 2000). Figure 5.11 shows effect of pH on denitrification rates (US EPA, 1993). The minimum COD/TN and alkalinity/TN ratios required for higher nitrogen removal are about 6 and 3.6, from Figures 5.10 and Figure 5.5, respectively.

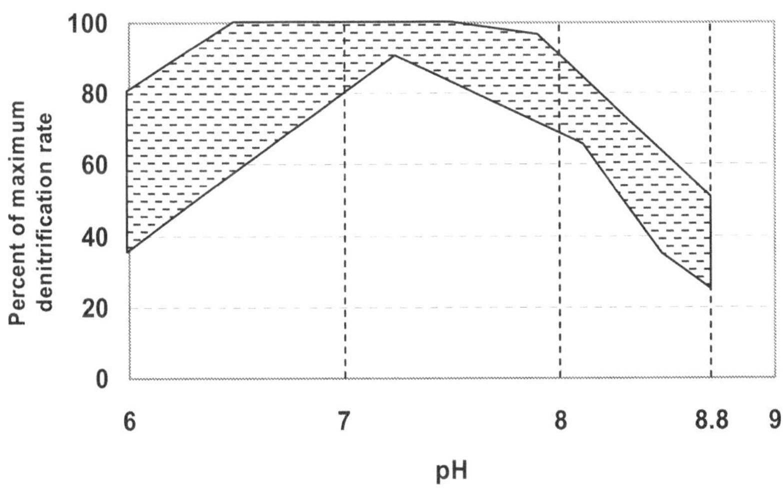

Figure 5.11 Effect of pH on denitrification rates for a sewage treatment plant (modified from US EPA, 1993).

(3) Denitrification Rates

The denitrification rate can be expressed similar to oxygen utilization in an aerobic systems :

$$\text{NUR} = \eta \frac{OUR}{2.86} \quad \text{for } NO_3\text{-N denitrification} \tag{5.16}$$

$$\text{NUR} = \eta \frac{OUR}{1.71} \quad \text{for } NO_2\text{-N denifrification} \tag{5.17}$$

$$\text{OUR} = \text{OUR}_{syn} + \text{OUR}_{endo} \tag{5.18}$$

$$OUR_{syn} = (1\text{-}Yn)\ \Delta COD/HRT \tag{5.19}$$

$$OUR_{endo} = (1\text{-}f_p)\ \Delta COD \left[\frac{Y_H}{1+b_H SRT}\right]\frac{SRT}{HRT}\cdot b_H \tag{5.20}$$

where

NUR = NO_x-N uptake rate, mg NO_x-N/L/hr
η = Correction factor
OUR = Oxygen uptake rate, mgO_2/L/hr
OUR_{syn} = OUR for synthesis
OUR_{endo} = OUR for endogenous
ΔCOD = Biodegradable COD removed/d
f_p = Inert particulate products from biomass decay, 0.2
Y_H = Heterotrophic yield, active mass
SRT = Solid retention time
HRT = Hydraulic retention time
b_H = Endogenous respiration rate based on active mass, Y_H

Figures 5.12 and 5.13 show a flow diagram of a full scale treatment plant and its performance data. This plant is designed for biological nitrification and denitrification followed by chemical coagulation and sand/carbon filtration. Figure 5.13 shows that the nitrogen removal efficiencies are strongly dependent on the operating temperature. Temperatures above 35°C have a tendency to reduce the removal efficiencies due to nitrification difficulties as previously discussed. The operating results of this plant will be discussed more in Chapter 7. Design information for this plant are as follows:

HRT = 10.7 day for anoxic reactor and 22.5 day for oxic reactor.

SRT $\geq$ HRT

From Equations (5.18) and (5.19)

$$OUR_{syn} = \frac{0.33\Delta COD}{10.7d} = 0.03\ \Delta COD$$

$$OUR_{endo} = 0.8\ \Delta COD\ [\frac{0.67}{1+0.48\times 33.2}]\times\frac{SRT}{HRT}\times 0.48 = 0.047\Delta COD$$

(Assumed SRT = total HRT, total HRT = oxic + anoxic HRT = 33.2 d)

ΔCOD = COD removed from biological process; 58.9 – 2.28 = 56.62 g/L/d

OUR = (0.03 + 0.047) ΔCOD=4,360 mg/L/d = 182 mgO_2/L/hr

Assuming NO_3-N denitrification prevails,

$$\text{NUR} = \frac{182}{2.86} = 63.6 \text{ mg NOx} - \text{N} / \text{L} / \text{hr}$$

This NUR is computed without consideration of a correction factor.

$$\text{Actual NUR} = \frac{\text{inf TKN} - \text{eff}(\text{TKN} - \text{NH}_4\text{N} - \text{NO}_x\text{N})}{\text{Anoxic HRT}}$$

$$= \frac{6850 - 921 - 33 - 65}{10.7 \times 24} = 22.7 \text{ mg NOx} - \text{N} / \text{L}$$

This assumes that all denitrification takes place in an anoxic reactor.

In addition, the difference is assumed to be due to not accounting the amount of organic nitrogen removed by soils and ammonia stripped, and not applying a correction factor. Choi and Rhu (2001) proposed factors of 1.0 for VFACOD, 0.5 for particulate COD and 0.3 for endogenous respiration. VFACOD is only 20% of the influent total COD. It can be assumed that a total correction factor of 0.6 can be applied where

OUR = (0.03×0.6+0.047×0.3) ΔCOD = 0.032 ΔCOD = 75.5 mg/L/hr

$$\text{NUR} = \frac{75.5}{2.86} = 26.4 \text{ mg/L/hr}$$

Figure 5.14 shows AURs and NURs measured from laboratory and full scale plants of piggery wastewater as well as other strong nitrogenous wastes such as nightsoil (Choi *et al.,* 1997) and human urine (Udert *et al.,* 2002). The AURs decrease as the COD/N ratios increase, while the NURs have a tendency to increase with the COD/N ratios.

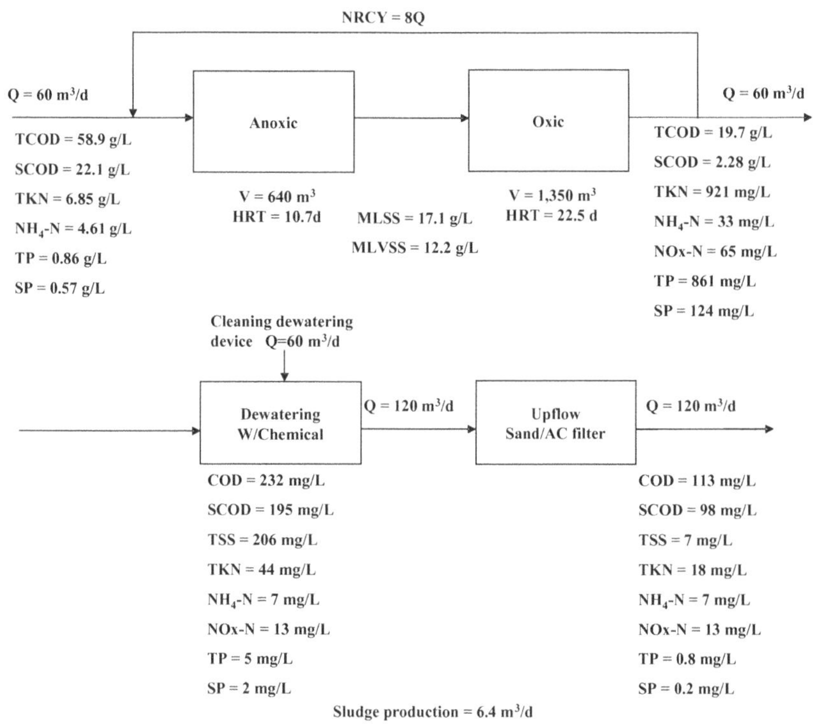

Figure 5.12 Flow diagram and values for process parameters of a full scale BNR treatment plant with enhanced coagulation and filtration processes (modified from Hwang and Min, 2006).

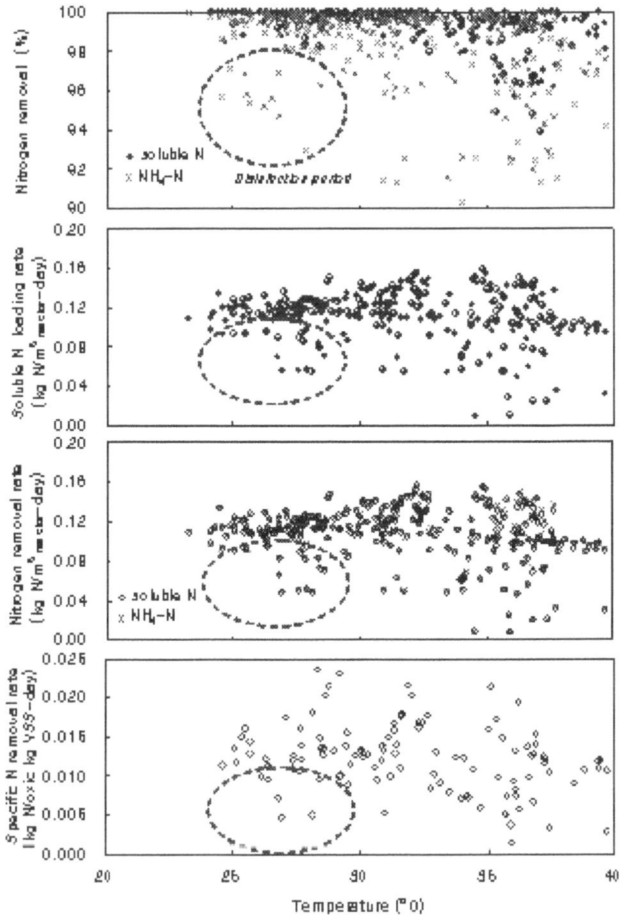

Figure 5.13 Nitrogen removal efficiencies, nitrogen loads and nitrogen removal rates with temperature variations in a full scale BNR plant treating 60 m^3/d of piggery wastewater (Hwang and Min, 2006). The data points in the dotted circle are presumed to be inhibited by pharmaceuticals in the influent during the spring.

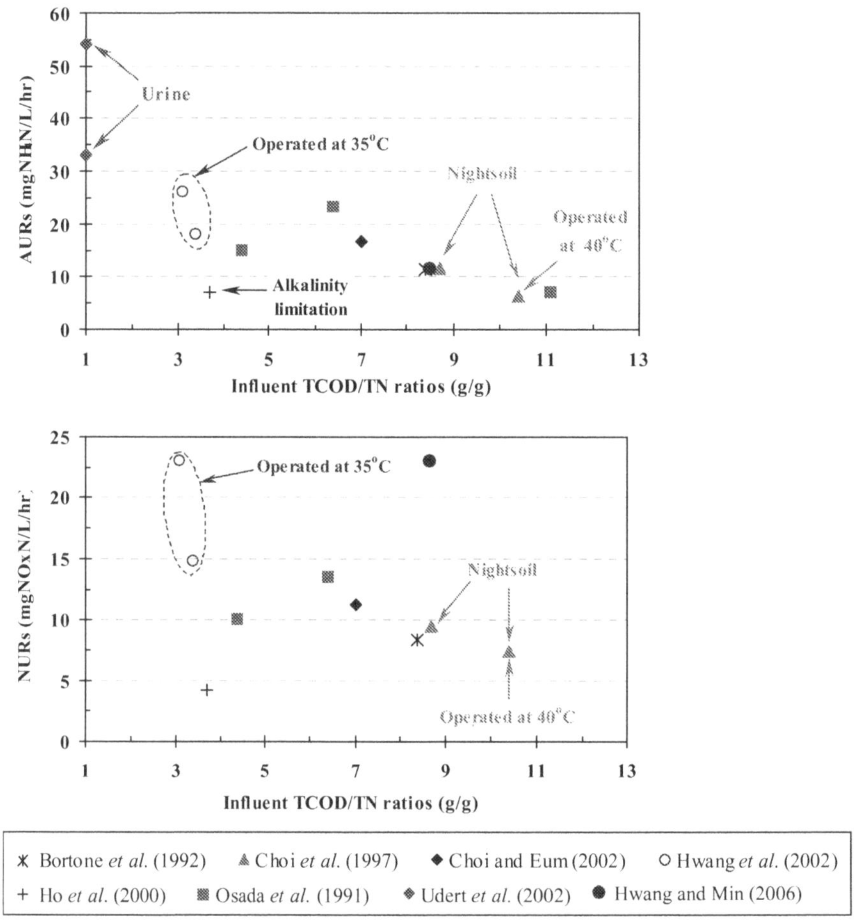

Figure 5.14 AURs and NURs with influent COD/TN ratios.

5.2.3 N_2O Emission

It has been reported that a large amount of N_2O is emitted from swine wastewater treatment plants (Osada *et al.,* 1991). The global warming potential of N_2O per molecule is about 15 to 300 times greater than that of CO_2 (IPCC 2001). Recently Kishida *et al.,* (2004) reported that N_2O emission was strongly dependent on the COD/N ratios in the denitrification process. The total N_2O emission for the operating condition of a 2.6 BOD/TN ratio was approximately

270 times greater than a 4.5 BOD/TN ratio. However, the effect of the COD/N ratio on N_2O emission from the nitrification process was not significant.

Hanaki *et al.,* (1992) observed N_2O emissions from nightsoil plants. Further, Yun *et al.,*(2001) indicated that the emissions increased at lower COD/TN ratios. This suggests that piggery plants with lower strengths would have a greater probability to emit more N_2O.

5.2.4 Nitrification and Denitrification Model

As shown in Figure 5.15, there occurs a variety of physical, chemical and biological reactions including NH_3 stripping, nitrite and nitrate nitrification and denitrification, chemical precipitation and resolubilization. The latter process can not be easily observed in sewage treatment. Piggery waste is introduced in an anoxic (Ax) condition (see Figure 5.1) where denitrification occurs with accompanying NOxN decreases at a negligible DO level. NH_4-N can be absorbed during heterotrophic cellular formation. Chemical precipitation also takes place with the formation of struvite and apatite as the pH increases. At oxic conditions, which follow the anoxic conditions, nitrification will take place along with NH_3 stripping due to aeration and redissolution (resolubilization) of the chemical precipitates due to the pH decrease. Organics in the feed are mostly removed in the anoxic stage and heat released during this stage. Aeration during the oxic stage also increases the reactor temperature with heated air from blowers as well as from the decomposition of the remaining organics. As discussed earlier, nitrite nitrification is beneficial for nitrogen removal from piggery waste, thus nitrogen loads must be higher. Consequently a higher organic load must be introduced resulting in higher reactor temperature. Typical design temperatures are 20 to 25°C during the winter and 35 to 40°C during the summer in Korea and the northeastern part of China.

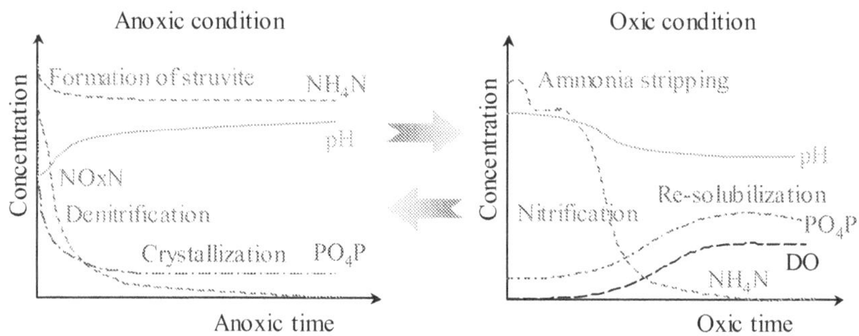

Figure 5.15 pH, DO, nitrogen and phosphorus profiles in a BNR bioreactor.

Figure 5.16 illustrates the transformations and interactions of carbon, nitrogen, phosphorus and oxygen in the anoxic and oxic stages including pH and alkalinity changes. COD is used for denitrification and alkalinity is recovered resulting in the pH increase in the anoxic stage. This results in the chemical precipitation of struvite and apatite. In the oxic stage, the remaining COD is removed by ordinary heterotrophic organisms (OHO) and nitrification will occur resulting in alkalinity and pH decreases. Some NH_3 will be stripped and the precipitated struvite and apatite will be dissolved.

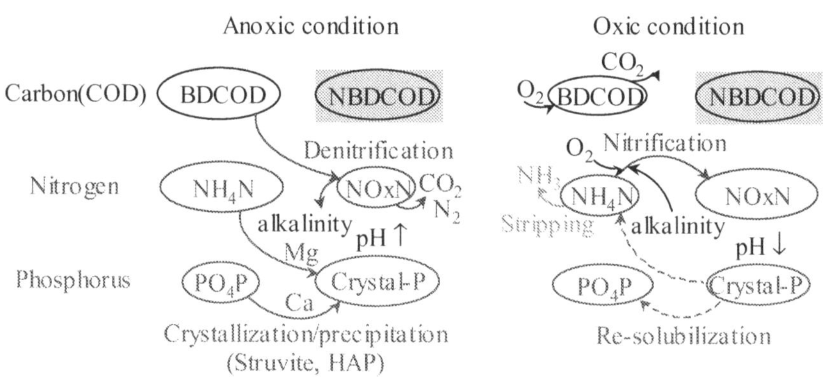

Figure 5.16 Transformation of COD, N, P and O along with changes of pH and alkalinity resulting from the biological reactions in anoxic and oxic stages of the BNR process for piggery waste.

5.3 KINETIC MODEL APPLIED FOR NITROGEN REMOVAL FROM PIGGERY WASTEWATER

(1) Nitrogen to be nitrified

$$N_i = \Delta N\ nitrified = TKN - 0.03(\Delta COD) - Stripped\ N - Struvite\ N - inert\ N \qquad \text{............ (5.21)}$$

$$Stripped\ N = \frac{\inf.Alk - eff.Alk - 7.1(\Delta N\ nitrified) + 3.6(\Delta N\ denitrified)}{7.1} \qquad \text{...........(5.22)}$$

Assuming 4% of nitrogen is stripped and 1% of nitrogen is precipitated during biological nitrification and denitrification, the amount of nitrogen to be nitrified can be computed from an alkalinity balance. The term of 0.03 (Δ COD) is the amount of nitrogen required for cellular formation.

(2) Required Ox HRT

Assume the effluent NH_4-N concentration is about 30 mg/L and determine the required HRT using Equation (5.9).

$$K_{25} = 5/d\ (at\ 25^{o}C)$$

$$eff.NH_4N = \frac{N_i}{K_{mn} \times HRT_{Ox} + 1} \qquad HRT_{Ox} = \frac{(N_i / eff.NH_4N) - 1}{K_{mn}}$$

(3) Anoxic HRT

Assume an anoxic HRT using the NUR data from Figure 5.14 considering the influent COD/ TKN ratio. Check the anoxic HRT with the OUR_{syn} and OUR_{endo} and determine the required anoxic HRT.

(4) OUR

Use Equation (5.18):

$$OUR = OUR_{syn} + OUR_{endo}$$

(5) NUR

$$NUR = OUR \times \frac{\eta}{\chi} \qquad (\chi = 1.71 - 2.86)$$

Apply the correlation factors considering the influent characteristics.

(6) Determine the total HRT

$$HRT_{Ax} = \frac{(N_i - eff.\,NH_4N)}{NUR} \qquad HRT_{total} = HRT_{Ox} + HRT_{Ax}$$

(7) Determine the COD removed in the anoxic reactor and determine the NRCY

Use Table 5.2:

$$\Delta COD_{Ax} = \frac{(COD / NO_x N) \times (\Delta N\ denitrified)}{1 - Y_{obs}}$$

$$\Delta N\ denitrified = \frac{(\Delta Alk\ recovered)}{3.6}$$

$$NOxN_{effluent} = \frac{Influent\ NOxN}{(RAS + NRCY + 1)}$$

(8) Computation of effluent COD

The biodegradable COD is almost removed when complete nitrification is accomplished. Thus, the effluent COD includes the nonbiodegradable COD and the unsettled biodegradable COD.

$$\text{Eff. COD} = \text{NBDCOD} + \text{O}_x\ \text{eff. BDCOD (soluble)} + 1.42\ X_{H\ eff}$$

$X_{H\ eff}$ is the unsettled active VSS in the effluent (effluent VSS × X_H/ total MLSS).

Figure 5.7 shows the overall results of how the influent concentrations vary with ambient temperatures and how this affects the effluent qualities in terms of COD, nitrogen and phosphorus. The basic data for simulations are as follows:

Basic data for simulation;

Influent COD = 20,000 mg/L, TKN = 3,200 mg/L at 25°C during winter

Influent COD = 10,000 mg/L, TKN = 1,700 mg/L at 35°C during summer

Based on winter operation, the required anoxic and oxic HRTs are 6 days each or a total of 12 days compared with the predicted total SRT of 36 days using the IWA ASM model (2000). Thus, the HRTs will be shortened to 3.3 day each with an increased flow which will be about 1.8 times the winter flow in Korea.

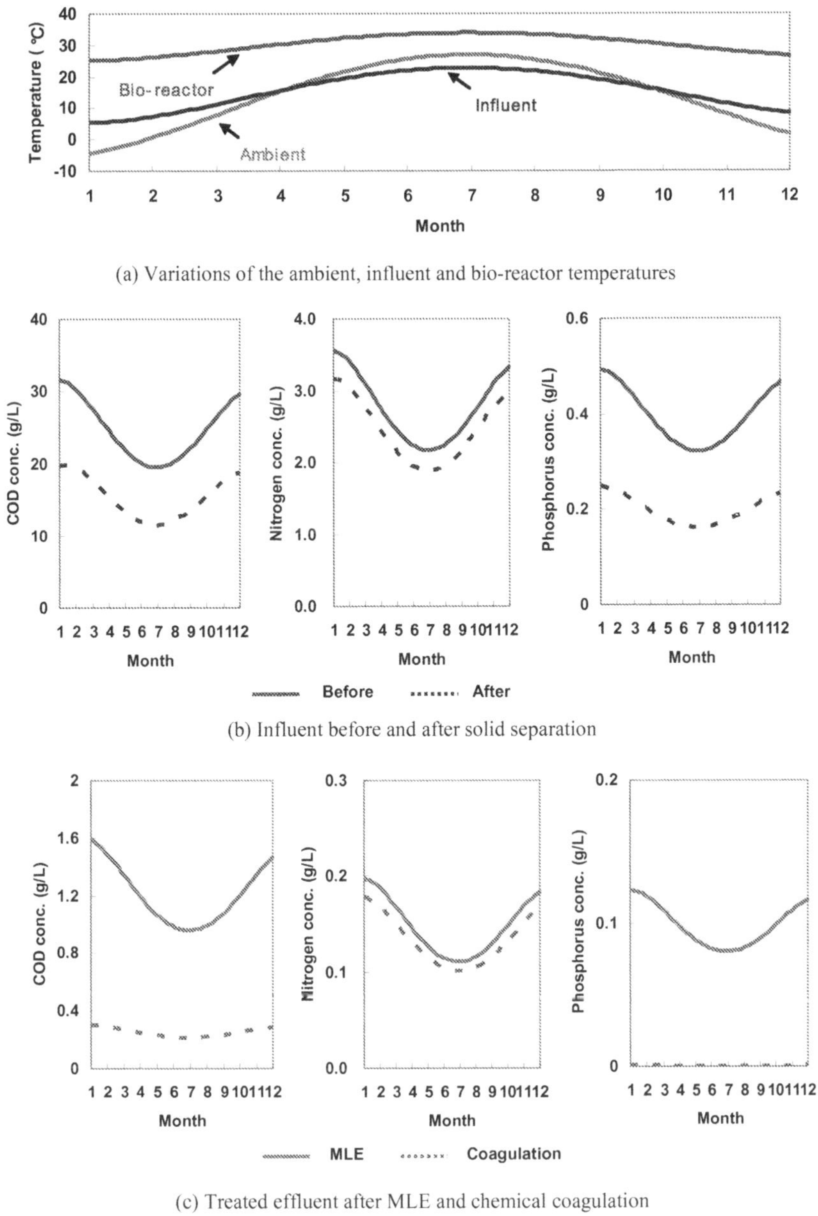

(a) Variations of the ambient, influent and bio-reactor temperatures

(b) Influent before and after solid separation

(c) Treated effluent after MLE and chemical coagulation

Figure 5.17 Predicted monthly variations of temperature and water quality changes before and after treatment.

5.4 ANAMMOX PROCESS

Mulder *et al.* (1995) found that ammonium can be converted into nitrogen gas using nitrite as the electron acceptor under anaerobic conditions in comparison to the conventional nitrification/denitrification process. The ANAMMOX (anaerobic ammonium oxidation) process when combined with the SHARON process has these advantages: organic carbon is not required for denitrification and a smaller amount of sludge is produced.

The SHARON process is a nitrite nitrification process where a higher nitrogen load is used to achieve a NH_4-N:NO_2-N ratio of 50:50 at a precisely controlled temperature and where different growth rates are achieved between the NH_4-N and NO_2-N oxidizers as illustrated in Figure 5.2. The SHARON and ANAMMOX equations can be expressed as follows:

$$\text{(SHARON)} \quad 2NH_4^+ + 1.5\ O_2 \rightarrow NO_2^- + 2\ H^+ + H_2O$$

$$\text{(ANAMMOX)}\ NH_4^+ + 1.32\ NO_2^- + 0.066\ HCO_3^- + 0.13\ H^+ \rightarrow 1.02\ N_2 + 0.26\ NO_3^- + 0.066\ CH_2O_{0.5}N_{0.15} + 2.03\ H_2O$$

The combined SHARON-ANAMMOX process is represented as Equation (5.23).

$$2\ NH_4^+ + 2\ HCO_3^- + 1.5\ O_2 \rightarrow N_2 + 2\ CO_2 + 5\ H_2O \qquad (5.23)$$

Piggery wastes generally show high organic and nitrogen concentrations. Thus, methane production and recovery processes are commonly used to reduce COD; however, it is difficult to do this if nitrogen removal is practiced since ordinary denitrification requires carbon which has already been removed in the previous methane generation step. Therefore, there have been some efforts to examine the applicability of ANAMMOX process for piggery waste (Hwang and Min, 2006, Min *et al.*, 2002; Ahn *et al.*, 2004; Hwang *et al.*, 2004, 2005). Anaerobic granules obtained from a brewery waste treatment plant were used to develop ANAMMOX bacteria for piggery waste.

A laboratory study with UASB, SHARON and ANAMMOX reactors was conducted with a piggery waste at 35°C and the results of water quality changes in terms of nitrogen, phosphorus and COD are shown in Table 5.3.

Table 5.3 Changes of nitrogen, phosphorus and COD concentrations for different processes based on an influent of 400 mL/day or 1,100 mL/d after elutriation in an acidogenic reactor with piggery waste (Hwang et al., 2005).

Description	TKN	NH_4-N	T-P	PO_4-P	COD	pH
Influent (400 mL/d)						
(mg/L)	5,730±780	4,300±60	1,770±300	440±50	58.600	8.5
(g/d)	2.29±0.31	1.70±0.02	0.71±0.12	0.18±0.02	23.4	
Acidogenic reactor						
Elutriated effluent (1,100 mL/d)						
(mg/L)	1,480±210	1,330±240	514±390	214±142	20,120	
(g/d)	1.77±0.25	1.60±0.27	0.62±0.47	0.26±0.17	24.14	7.3
Wasted sludge (150 mL/d)						
(mg/L)	1,690±300	1,170±260	910±250	166±70	35,630	
(g/d)	0.15±0.05	0.12±0.03	0.08±0.03	0.008±0.007	3.11	
Methanogenic UASB reactor						
effluent (1,100 mL/d)						
(mg/L)	1,310±160	960±160	204±105	114±140	4,540	8.1
(g/d)	1.44±0.18	1.06±0.17	0.22±0.12	0.13±0.02	4.53	
SHARON reactor						
effluent (1,100 mL/d)						
(mg/L)	564±105	225±49	147±36	51±15	3,050	8.3
(g/d as 1,100 mL/d)	0.62±0.12	0.25±0.05	0.16±0.04	0.06±0.02	3.36	
ANAMMOX reactor						
effluent (1,100 mL/d)						
(mg/L)	285±99	92±14	86±28	46±4	1,350	8.7
(g/d as 1,100 mL/d)	0.31±0.11	0.10±0.02	0.09±0.03	0.05±0.004	1.49	
Removal based on effluent (%)	86.5	94.1	87.3	72.2	93.6	-

Table 5.4 compares ANAMMOX treatment of piggery waste with reject water from a municipal plant. The piggery waste was fermented with a 2 day HRT including elutriated water which increased the influent flow rate, by 2.5 times and the fermented liquor was introduced to UASB reactor operated with a 2 day

HRT (or 5 day HRT based on the influent flow rate). The UASB reactor removed 88% of the COD and the effluent from the UASB reactor was introduced to the SHARON reactor operated with 2.5 day HRT followed by an ANAMMOX reactor operated with 6.25 day HRT based on the influent flow rate. Influent concentrations were 58,000 mg/L COD, 5,700 mg/L TKN (4,300 mg/L NH_4-N) and 1,770 mg/L TP. About 400 mg/L NO_2-N and 250 mg/L NH_4-N were introduced to the ANAMMOX reactor and about 200 mg/L NO_2-N and 100 mg/L NH_4-N were discharged. The NH_4-N removal efficiency was about 60% in the ANAMMOX reactor, but the overall nitrogen removal efficiency reached 90% with CH_4 gas recovery. About 40% of the total phosphorus (TP) removal was accomplished in the ANAMMOX reactor due to the increase of pH, and the overall TP removal efficiency with this system including fermentation, CH_4 gas recovery and the SHARON reactor was 88%.

Table 5.4 Summary of operating characteristics for ANAMMOX reactors treating municipal reject water and piggery waste.

Items	Reject water (Mulder *et al.*, 1995)	Piggery waste (Hwang *et al.*, 2005)
Feed	Reject water produced from municipal plant after anaerobic sludge digestion with little BD* COD remaining after nitrification	Piggery waste treated with anaerobic digestion with little BD COD remaining after nitrification
Seeding	ANAMMOX biomass obtained from various sources	Anaerobic granules obtained from brewery waste plant
Concentrations (mg/L)	COD = little BD COD remaining NH_4-N = 550 NO_2-N = 600 NO_3-N = 0 NO_2-N/NH_4-N = 1~1.3	COD = 2,460 (total) with little BD COD remaining NH_4-N = 213 NO_2-N = 323 NO_3-N = 144 NO_2-N/NH_4-N = 1.56
Applied N loads (kgN/m^3-d)	1.2	1.7 1.36 with soluble N
HRTs (d)	1	6.25
Time required for steady states (d)	About 100	About 26
N removal rates	0.91~0.96 kg N/m^3-day 0.70 kg N/kg TS-day	0.72 kg soluble N/m^3-day 0.1 kg soluble N/kg VSS-day
NO_2-N:NH_4-N ratio	1.2	0.6 to 2.13

*BD means biodegradable.

5.5 PHOSPHORUS REMOVAL IN BNR SYSTEM

Biological nutrient removal for municipal wastewater has become a popular process which also includes phosphorus removal. The process includes three stages; anaerobic, anoxic and oxic stages. Both PAO (phosphorus accumulating organisms) and denitrifiers compete for VFA COD which includes very soluble forms of COD. Minimum dissolved oxygen (DO) and NO_x-N concentrations are desirable in the return activated sludge (RAS). PAO store VFA COD in their cellular mass and releases PO_4-P into the liquid in the anaerobic stage. The released PO_4-P is taken up by the PAO in the oxic stage resulting in phosphorus removal. NH_4-N is oxidized in oxic stage and the nitrified liquor is recycled back to anoxic stage to be denitrified.

Piggery wastes consist of high organics and nitrogen. The organics are the carbon source for PAO and denitriers to facilitate the removal of the phosphorus and nitrogen. A minimum VFA COD requirement for P and NO_x-N removal from sewage is about 5 to 10 g COD/g P and 2 to 5 g COD/NO_x-N when the COD is not consumed by other hetrotrophs (Choi *et al.*, 2001a). Practically, phosphorus removal is not easy with piggery wastewater, since COD is normally removed by anaerobic digestion which does not remove nitrogen resulting in lower nitrogen and phosphorus removal efficiencies in BNR systems due to the lower COD/N ratios. However, it was observed that about 70% phosphorus was removed by chemical precipitation with struvite and hydroxyapatite (HAP) and cellular formation. This is basically related to pH along with organic and nitrogen loads that resulted in influent COD/N ratios as shown in Figure 5.18. The removal efficiencies increased from 50 to 90% as COD/N values increased from 6 to 7, but carbon was not limited beyond these values for denitrification resulting in a stable pH. Figure 5.19 shows phosphorus profiles for various treatment or resource recovery processes. Phosphorus is not removed in the oxic stage because of a decrease in the pH due to nitrification. In order to maximize phosphorus removal in a BNR system, an additional anoxic stage must be furnished prior to the discharge of its final effluent after the initial oxic stage. Phosphorus contents of the BNR sludge was about 7.1 to 8.5% of which cellular P was about 2% as shown in Figure 5.20 (Choi *et al.*, 2006).

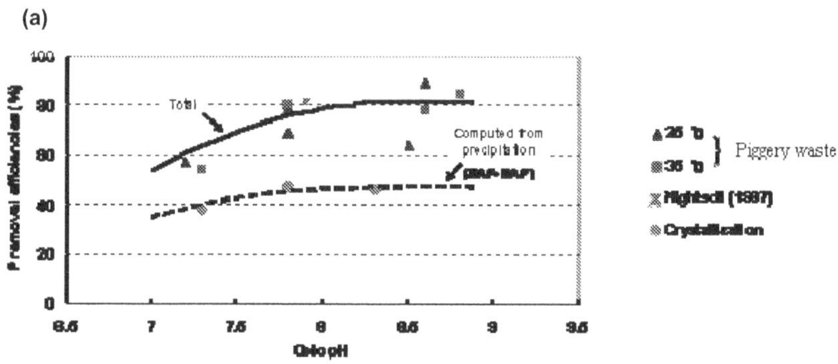

(a) Oxic pH

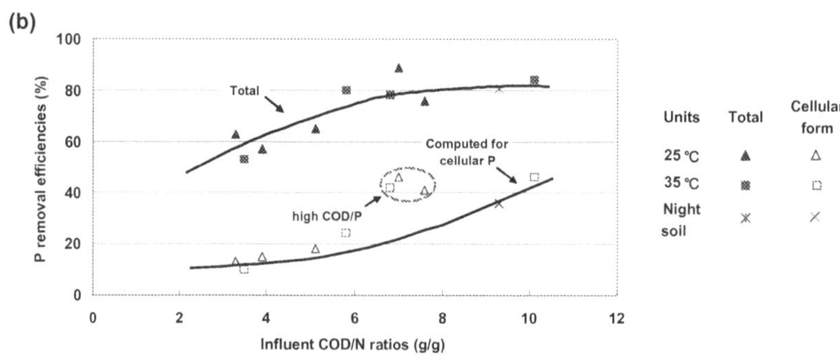

(b) COD/TKN ratios

Figure 5.18 Phosphorus removal efficiencies with oxic pH and influent COD:TKN ratios in the BNR process.

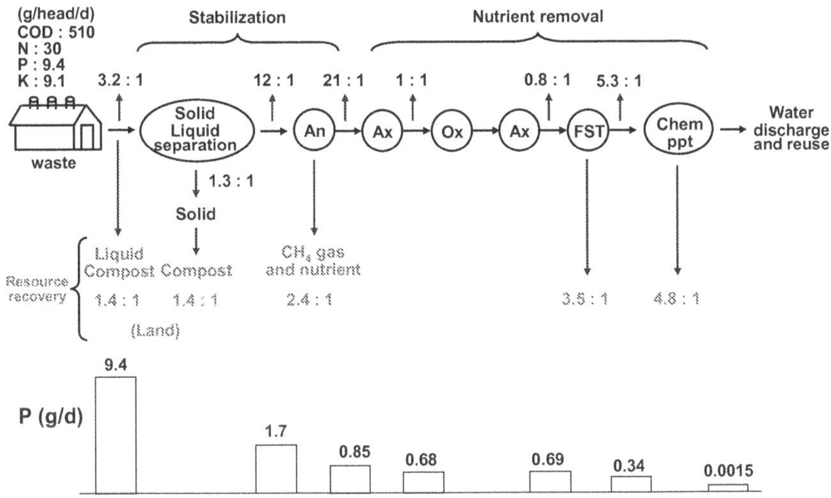

Figure 5.19 N and P profiles for different piggery waste treatment processes.

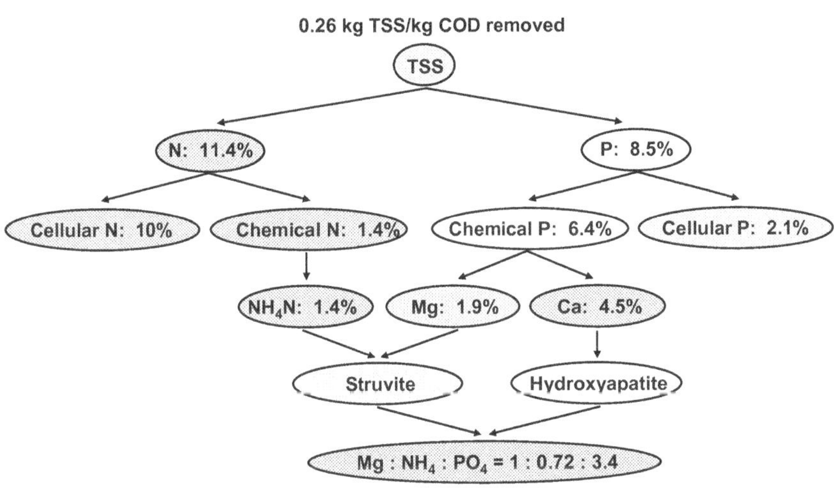

Figure 5.20 Fractionated P contents of oxic sludge including the struvite and hydroxyapatite contents of the wasted sludge from the BNR process at 35°C The waste sludge represents 8.5% P of total sludge.

REFERENCES

Ahn, Y H., Hwang, I S. and Min, K S (2004) ANAMMOX and partial denitrification in anaerobic nitrogen removal from piggery waste. *Wat Sci. Technol.*, **49**(5-6), 145-153.

Anthonisen A C, Loehr R C, Prakasam T B S and Srinath E G (1976) Inhibition of nitrification by ammonia and nitrous acid, *JWCF*, 48, 835-842.

ASM model (2000) *IWA task group on mathematical modeling for design and operation of biological wastewater treatment*, IWA Publishing.

Beline F, Boursier H, Guiziou F and Paul E (2007) Modelling of biological nitrogen removal during treatment of piggery wastewater, *Wat. Sci. Technol.* **55**(10), 11-19.

Bicudo J R, Svoboda I F (1995) Effect of intermittent-cycle extended-aeration treatment on the fate of carbonaceous material in pig slurry, *Bioresource Technology,* 54, 53-62.

Bortone G, Gemelli S, Rambaldi A and Ticlcyee A (1992) nitrification, denitrification and biological phosphate removal in sequencing batch reactors treating piggery wastewater, *Wat. Sci. Technol.* **26**(5), 977-984.

Choi E and Eum Y (2001) Strategy for nitrogen removal from piggery waste by nitrite nitrification. *Wat. Sci. Technol.* **44**(6-7), 347-353.

Choi E, Park H and Rhu D (2001a) Phosphorus removal from SBR with controlled denitrification for weak sewage, *Wat. Sci. Technol.* **43**(3), 159-166.

Choi E and Rhu D (2001) NUR and OUR relations in BNR process with sewage at different temperature and its design application, *Wat. Res.* **37**(7), 1748-1755.

Choi E. Eum Y, Gil K-I and Oa S W (2004) High strength nitrogen removal from nightsoil and piggery wastes, *Wat. Sci. Technol.* **49**(5-6), 97-104.

Choi E, Kim D, Eum Y, Yun Z and Min K-S (2005) Full-scale experience for nitrogen removal from piggery waste, *Wat. Env. Res.* **77**(4), 381-388.

Choi E, Oa S W and Lee J-J (1997) Nightsoil treatment plant converted into sequencing batch reactor to improve removal of pollutants and nutrients, *Wat. Sci. Technol.* **35**(1), 233-240.

Choi E, Y B Yu, M C Cui, Z Yun and K S Min (2006) Effect of biologically mediated pH change on phosphorus removal in BNR system for piggery waste treatment, *Proc of the 7th IWA symposium on waste management problems in agro-industries*, 191, Amsterdam.

Choi E, Zhou L, Ha J S and Hwang H-J (2003) Nightsoil treatment for China: present and future perspectives, *Water in China* edited by Wilderer P A, Zhu J and Schwarzenbeck N, 171, IWA Publishing

Eum, Y (1997). *Nutrient removal from settled piggery wastewater with low COD/N ratio*, MS thesis, Department of Civil & Environmental Engineering, Korea University, Seoul, Korea.

Eum Y and Choi E (2002) Optimization of nitrogen removal from piggery waste by nitrite nitrification, *Wat. Sci. Technol.* **45**(12), 89-96.

Eum Y, Oa S W and Choi E (2003) Application of piggery waste to nightsoil plant towards sustaianable development, *Wat. Sci. Technol.* **47**(10), 147-154.

Gil K, Choi E, Yun Z and Lee J-W (2002) The nomographic design approach of recycle water treatment by the nitritation process, *Wat. Sci. Technol.* **40**(11-12), 85-92.

Hanaki K, Zheng H and Matsuo T (1992) Production of nitrous oxide gas during denitrification, wastewater, *Wat. Sci. Technol.* **26**(5-6), 1027-1034.

Ho J H and Choi E (2000) Nitrogen removal from piggery waste by nitrite nitrification, *Proc of WEFTEC 2000*, Anaheim, CA, USA, Session 30.

Howarth R W (2004) Human acceleration of the nitrogen cycle: drivers, consequences, and steps toward solutions, *Wat. Sci. Technol.* **49**(5-6), 7-14.

Hwang B, Hwang K Y, Choi E, Choi D K and Jung J Y (2000) Enhance nitrite build-up in proportion to increasing alkalinity/NH4 ratio of influent in biofilm reactor, *Biotech letters*, 22, 1287-1296.

Hwang H J, Eum Y, Choi E (2002) nutrient removal from piggery waste with SBR coupled with membrane: A full scale experience. *Proc of WEFTEC 2002*, Chicago, IL.

Hwang, I S, Min, K S (2006) Piggery waste treatment using improved MLE process full sacle, *J KSWQ*, 22, 5, 895-904.

Hwang, I S., Min, K S., Choi, E and Yun, Z (2005) Nitrogen removal from piggery waste using the combined SHARON and ANAMMOX reactor. *Wat. Sci. Technol.* **52**(10-11), 487-494.

Hwang, I.S., Min, K.S. and Lee, Y.O. (2004) Nitrogen removal from piggery wastewater using ANAMMOX reactor. *Proc. 10th World Congress of Anaerobic Digestion*, 29 Aug. - 2 Sep., Montreal, Canada, 1, 261-266.

IPCC (2001*) Climate change 2001: The scientific basis*, Cambridge university press, Cambridge, UK

Jenicek P, Svehla P, Zabranska J and Dohanyos M. (2004) Factors affecting nitrogen removal by nitritation/denitrification, *Wat. Sci. Technol.* **49**(5-6), 73-79.

Kishida N, Kim J H, Kimochi Y, Nishimura O, Sasaki H and Sudo R (2004) Effect of C/N ratio on nitrous oxide emission from swine wastewater treatment process, *Wat. Sci. Technol.* **49**(5-6), 359-366.

Magri A, Flotats X (2000) Biological treatment of the liquid fraction of pig slurry in a sequencing batch reactor, *2and International Symposium on SBR,* Vol.2, Poster Presentations.

McCarty P L, Beck L and Amant P (1969) biological dentrification of wastewater by addition of organic materials, *Proc of the 24th Purdue Ind Waste conference*, Purdue university, Lafayette, IN.

Min, K S, Ahn, Y H, Hwang, I S and Choi, E (2002) Feasibility of ammonium removal from piggery waste by an anaerobic sludge bed reactor. *Proc. Animal Residuals Conference and Workshop*, May 6-8. Washington DC, U.S.A., 373-386.

Mulder A., van de Graaf A A, Robertson, LA and Kuenen JG (1995) Anaerobic ammonium oxidation discovered in a denitrifying fluidized bed reactor. *FEMS Microbiol., Ecol.*, 16, 177-183.

Osada T, Haga K, Harada Y (1991) Removal of nitrogen and phosphorus from swine wastewater by the activated sludge units with the intermittent aeration process, *Wat. Res.* **25**(11), 1377-1388.

Snoeyink V L and Jenkins D (1980) Water chemistry, John Wiley & Sons, Inc.

Tilche A, Bortone G, Malaspina F, Piccinini S, Stante L (2001) Biological nutrient removal in a full-scale SBR treating piggery wastewater results and modeling, *Wat. Sci. Technol.* **43**(3), 363-371.

Udert K M, Fux C, Munster M, Larsen T A, Siegrist H and Gujer W (2003) Nitrification and autotrophic denitrification of source separated urine, *Wat. Sci. Technol.* **48**(1), 119-126.

US EPA (1993) *Process design manual for nitrogen control,* EPA/625/R-93/010, Washington DC, USA.

Yu Y B (2007) The dynamic nature of MLE process for piggery waste treatment with temperature, MS thesis, Department of Civil and Environmental Engineering, Korea University, Seoul, Korea.

Yun Z, Hong Y, Lyou J, Choi E and Min K-S (2001) The evaluation of effects of system pH on nitrite denitrification with gas analysis, proc of 1st Asia Pacific regional conference, Fukuoka, Japan.

6

Post Treatment

6.1 INTRODUCTION

After biological treatment, there will still be COD remaining in the effluent. It will consist of mostly nonbiodegradable fractions with a dark brown color. In addition, there will be some of NH_4-N and NO_3-N remaining in the effluent. Depending on the existing or anticipated regulatory effluent standards, these impurities may require removal. For instance, Korean standards require for a centralized piggery plant that the final effluent be less than 200 mg/L COD, 30 mg/L SS, 60 mg/L TN and 8 mg/L TP. Chinese standards also requires a final effluent with less than 400 mg/L COD, 200 mg/L SS, 150 mg/L BOD, 80 mg/L NH_4-N, and 8 mg/L TP (Chinese regulation, 2001). About 10% of the organics discharged from pig houses is nonbiodegradable consisting of 4% in soluble and 6% in suspended forms (see Figure 3.6). With a proper solids separation, it is believed that 94 g COD, 20 g N and 1.7 g P per head per day is discharged with 5 to 8 L flow/head/d (see Figure 2.5). Thus, a typical effluent from a biological treatment plant would have a dark brown color with 900 to 1,600 mg/L COD, 80 to 120 mg/L SS, 20 to 100 mg/L NH_4-N, and 60 to 120 mg/L TP based on 5 to 8 L of flow. Table 6.1 shows influent and effluent qualities for the biological treatment of piggery wastewater. The influent is usually settled prior to application to the biological reactor. The effluent qualities vary with the influent concentrations which are affected by ambient temperature, since pigs drink more water during the summer and less water during the winter.

Table 6.1 Influent and effluent qualities from a piggery waste biological treatment process in Korea (mg/L).

Constituents	Winter		Summer	
	Influent	Effluent	Influent	Effluent
COD	20,000	1,600	10,000	900
BOD	10,000	Nil	5,000	Nil
TN	3,200	100 - 200	1,700	100-200
SS	4,000	80	2,500	120

Figure 6.1 presents particle size distributions for the effluent discharged from a piggery waste biological treatment plant (Lee *et al.,* 2004). The larger molecules probably originated from tannin and humic acids. Table 6.2 shows the measured tannin and humic acids contents expressed as COD. COD equivalents used for tannin acid and TOC for humic acid are 1.43 and 2.7, respectively. Tannin and humic acids are likely the major parts of the COD in biologically treated piggery wastewater effluents. Figure 6.2 shows the chemical structures of different types of tannin acids, whose chemical formula can be expressed as $C_{76}\,H_{52}\,O_{46}$ (IARC, 1976).

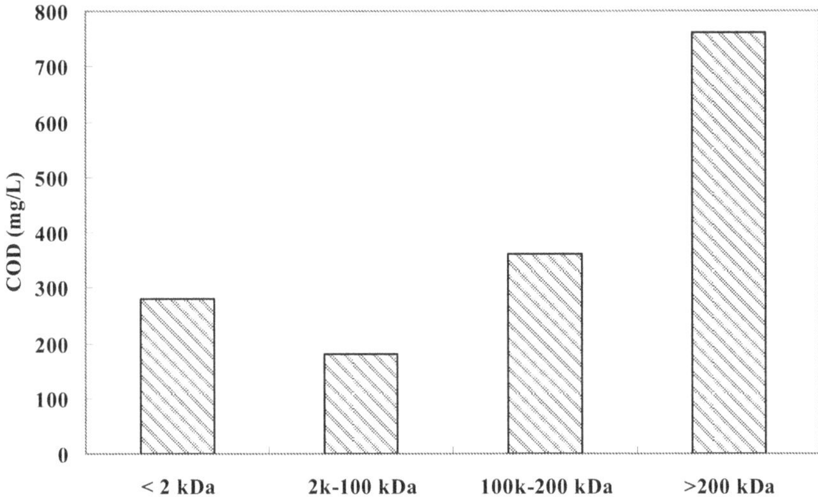

Figure 6.1 Effluent particle size distributions for a piggery wastewater from a centralized biological treatment plant. kDa denotes molecular weights in 1000Daltons.

Table 6.2 Tannin and humic acids contents in biologically treated piggery wastewater effluents (mg/L).

Bio- reactors	COD (mg/L)	Tannin acids (mg/L)		Humic acid (mg/L)	
		Tannin	COD*	Humic acid	Est. COD**
Lab at 25°C	1,320	680	972	107	289
and 35°C	1,250	620	886	93	251
Pilot	1,385	710	1,015	110	297
Full	992	500	715	92	248

* Estimated from $C_{76}H_{52}O_{46} + 66O_2 = 76CO_2 + 26H_2O$ and COD/tannin=1.43

** Estimated from COD/TOC = 2.7

Tannin and humic acids were measured by the 5550B colorimetric method and the 5510C XAD method, respectively, in accordance with *Standard Methods* (1998)

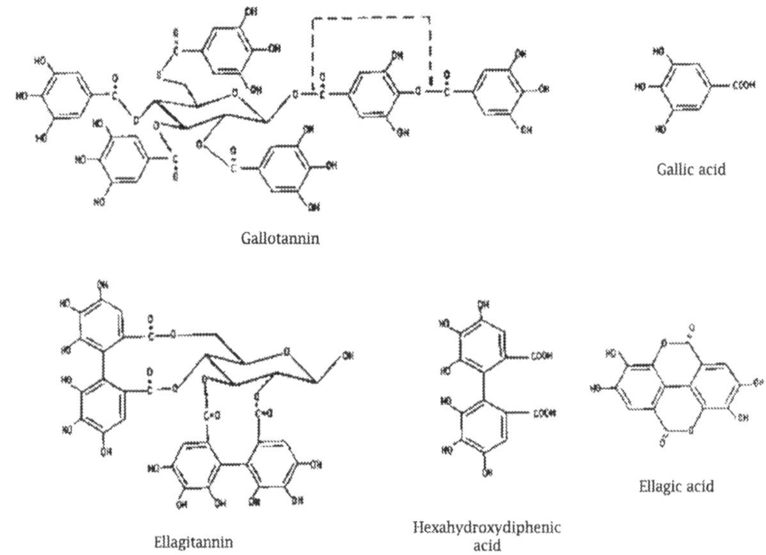

Figure 6.2 Chemical structures of tannin acids (IARC, 1976).

Table 6.3 compares the relative removal capabilities of various processes used to polish effluents from piggery wastewater biological treatment plants. The advanced oxidation process (AOP), chemical coagulation, carbon adsorption and membrane technologies are commonly used for the removal of COD and other constituents. To secure a complete nitrification, biofilm processes can be used in addition to suspended nitrification processes. Needless to say, the residual NO_x-N can further be removed by external carbon addition. Membrane treatment is a superior technology for separation of effluent solids, but only reverse osmosis (RO) is adequate to remove soluble forms of nonbiodegradable COD, NH_4-N and NO_x-N.

Table 6.3 Relative removal capabilities of processes for polishing piggery wastewater effluents from biological treatment plants.

	NBDCOD	NH4-N	NOx-N	PO4-P
AOP	XXX	X		
Chemical Coagulation	XXX			XXX
Adsorption	X	X	?	X
Membrane (RO)	XXXX	XX	XX	X

The greater the number of X's, the greater the relative removal capability.

6.2 CHEMICAL COAGULATION

Biologically treated piggery wastewater effluents consist of large molecules as shown in Figure 6.1. The large molecules have been effectively removed by chemical coagulation in water treatment plants (Jacangelo, 1995). Figure 6.3 shows how COD is removed with Fe^{3+}. COD is reduced from 1,600 to about 400 mg/L with chemical dosage of 300 mg/L resulting in about 4 kg COD removed/kg Fe^{3+} dosed. The sludge production is 0.2 m^3/m^3 waste without dewatering or 2.5 kg TS/kg Fe^{3+} dose. Actual plant operation indicates the dewatered sludge production is about 1% of the influent flow rate (Hwang and Min, 2006).

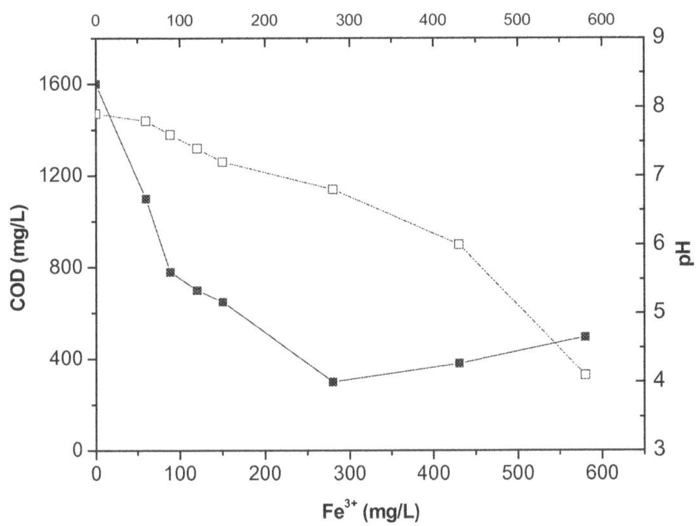

Figure 6.3 COD removal and pH changes with a Fe^{3+} coagulant dose (Choi, 2007). Jar test results were obtained with 1 min rapid mix and 10 min slow mix followed by 1 hr settling.

6.3 ADVANCED OXIDATION PROCESSES

Table 6.4 shows AOP available for water and wastewater treatment. The aim of the oxidation is to convert organic pollutants into simple and relatively harmless inorganic molecules. Biological oxidation is the mostly commonly applied technology for this purpose, but its application is limited to biodegradable organics. Various types of AOP have been used in laboratory and full scale units for the treatment of biologically treated piggery wastewater effluents and they include: Ozonation, Fenton oxidation, electrolysis and others.

Table 6.4 Advanced oxidation processes used for water and wastewater treatment.

Chemical oxidation	Ozone (O_3), hydrogen peroxide (H_2O_2), Cl_2
Physical processes, Irradiation	UV, UV/TiO_2, electron beam
Electrochemical process	Illustrated in Figure 6.5
Elevated conditions	Wet air, supercritical oxidation

6.3.1 Basic mechanism

The aim of most AOP is to produce free hydroxyl radicals in water. These free radicals are not ionic species, but are formed by photolysis from an equal cleavage with a two electron bond.

$$HO:OH \rightarrow HO^{*} + {}^{*}OH \quad \text{photolysis} \tag{6.1}$$

Each HO^{*} is uncharged and two HO^{*} will combine to form HOOH (also uncharged). The *symbol indicates the radical center and represents a single unpaired electron. Once a free radical reaction has been initiated by photolysis via ozonation, hydrogen peroxide, heat or other means, then a series of simple reactions will ensue. The complexity of the chemistry of such systems is due to the very large number of reactions that are possible. With such complicated mechanisms it is very difficult to predict all of the products resulting from the ensuing processes. The reaction rate will depend on the radical, oxygen and pollutant concentrations. Many factors can affect the radical concentration including pH, temperature, the presence of ions, the type of pollutant, as well as the presence of scavengers such as the bicarbonate ion, which is an effective radical trap (Parsons, 2004). Also, NO_2-N and NO_3-N are considered to be scavengers (Park *et al.,* 2006s; Park *et al.,* 2006f).

The radical trap by bicarbonate is illustrated in the following equations:

$$OH^{*} + HCO_3^{-} \rightarrow {}^{*}CO_3^{-} + H_2O \tag{6.2}$$
$$OH^{*} + CO_3^{2-} \rightarrow {}^{*}CO_3^{-} + OH \tag{6.3}$$

Figure 6.4 shows how the OH^{*} radical reacts with large molecules which consist of nonbiodegradable organics. The OH^{*} radical solubilizes the large molecules into smaller ones which can be degraded by microbes.

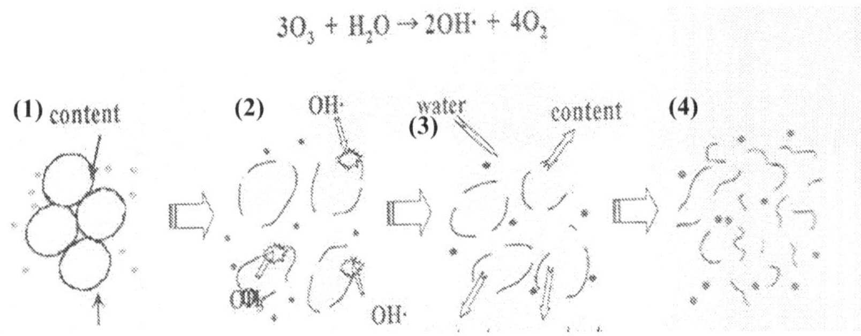

(1) large molecules in liquid
(2) and (3) molecules or microbial cells are solubilized by the OH^* radicals
(4) smaller molecules are degraded by microbes.

Figure 6.4 Hypothetical scheme of ozonation produced OH^* radicals disintegrating a large molecule.

The Fenton reagent is a mixture of H_2O_2 and ferrous iron, which generates OH^* radicals at an optimum pH of 3.5:

$$Fe^{2+} + H_2O_2 \rightarrow Fe^{3+} + OH^* + OH^- \quad (6.4)$$

The ferrous iron catalyses the decomposition of H_2O_2 resulting in the generation of hydroxyl radicals. The generation of the radicals involves a complex reaction sequence in an aqueous solution. Afterwards, the hydroxyl radical may either oxidize another Fe^{2+} to Fe^{3+}.

$$OH^* + Fe^{2+} \rightarrow OH^- + Fe^{3+} \quad (6.5)$$

or they may react with H_2O_2 as follows:

$$OH^* + H_2O_2 \rightarrow H_2O + HNO_2^* \quad (6.6)$$
$$HO_2^* + Fe^{3+} \rightarrow O_2 + Fe^{2+} + H^+ \quad (6.7)$$

Also, the hydroxyl radical may oxidize other species (substrate or RH) present in the solution:

$$OH^* + RH \rightarrow R^* + H_2O \rightarrow Products \quad (6.8)$$

The sequence of Reactions (6.4) and (6.5) corresponds to the Fe^{2+} stoichiometric oxidation by H_2O_2 (primary reaction). The sequence of Reactions (6.4) to (6.7) corresponds to the Fe^{2+} catalytic disproprotionation of H_2O_2.

Reactions (6.5) and (6.8) are in competition, so a relatively high Fe^{2+} concentration will favor the primary reaction and organic substances will not be oxidized, whereas relatively low Fe^{2+} concentration will lead to induced chain reactions where organic substances are oxidized. Reactions (6.4) to (6.7) correspond to the Fenton reactions. Reaction (6.8) which is a propagation step and the origin of an induced chain reaction where the substance will be finally oxidized (the Fenton reaction). Like Fenton process, all other AOP produce OH^* radicals which react with organics. When electrolysis is applied to piggery wastewater, a similar formation of OH^* radicals occurs as with chlorination shown in Table 6.4 and illustrated in Figure 6.5. It should be noted that denitrification can also be accomplished in these processes (Lee *et al.,* 2000s; Chae *et al.,* 2004). Also, bear in mind that Korean piggery wastewater contains about 1,600 mg/L Cl^-.

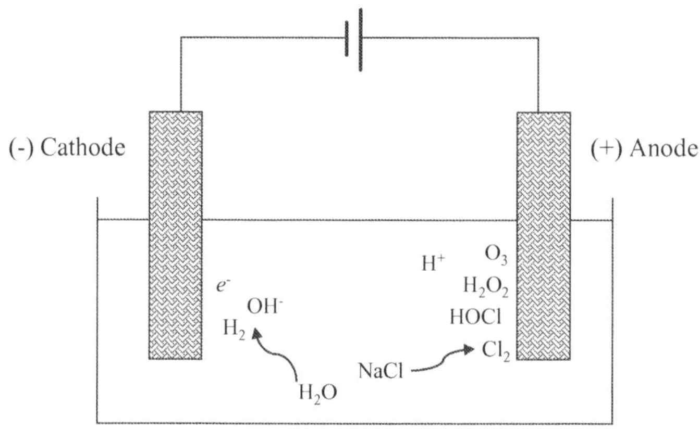

Figure 6.5 Electrolysis of piggery wastewater showing the formation of OH^* radicals.

Cathode reaction

$$2\,H_2O + 2\,e^- \rightarrow H_2\uparrow + 2\,OH^-$$
$$2\,NO_3^- + 6\,H_2 \rightarrow N_2\uparrow + 6\,H_2O$$

Anode reaction

$$2\,Cl^- \rightarrow Cl_2 + 2\,e$$
$$Cl_2 + H_2O \rightarrow HOCl + Cl^- + H^+ \quad \text{(Hydrolysis)}$$
$$HOCl + H_2O \rightarrow ClO_2 + 3\,H^+ + 3\,e$$
$$2\,H_2O \rightarrow O_2 + 4\,H + 4\,e^-$$
$$O_2 + H_2O \rightarrow O_3 + 2\,H^+ + 2\,e^-$$
$$2\,H_2O \rightarrow H_2O_2 + 2\,H^+ + 2\,e^-$$
$$O_3 + OH^- \rightarrow OH^*$$

6.3.2 Degree of Chemical Oxidation

Chemical oxidation in general includes three reactions: primary oxidation, acceptable oxidation, ultimate oxidation or fusing (Bowers, 1997). Primary oxidation involves a rearrangement of the original compounds to the extent that their chemical structures are altered and the oxidation byproducts may be more or less toxic than the original compounds. Acceptable oxidation involves the conversion of the original compounds to byproducts which have reduced toxicity. Ultimate oxidation involves the complete destruction of all organic species to CO_2 and H_2O as shown in Figure 6.6. Chemical oxidation is used for the tertiary treatment of biologically treated piggery wastewater effluents to achieve the complete destruction of residual nonbiodegradable organics.

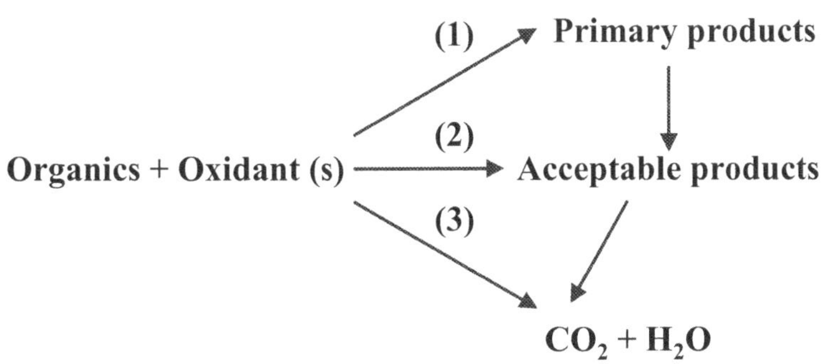

Figure 6.6 Chemical oxidation results in changes to chemical structures, reduced toxicity and production of a stable product.

6.3.3 Stoichiometry of Chemical Oxidation.

The reactions for all the different kinds of chemical oxidants may be compared based on the number (n) of free reactive oxygens (O* for oxygen) produced (Bowers, 1997). Ozonation produces one O* at a low pH and three O* at a high pH. H_2O_2 produces one O* as shown:

$$H_2O_2 \rightarrow O^* + H_2O \tag{6.9}$$

And the stochiometric oxidant dosage in molar units for COD may be computed from the molecular weight of oxidant/16 n times the COD.

Table 6.5 Chemical requirements for polishing biologically treated piggery wastewater effluents with chemical coagulation and oxidation processes.

Processes	Chemical coagulation		H_2O_2/Fe^{2+}		$H_2O_2/Fe^{2+}/UV$	O_3		
COD removed (mg/L)	1280	623	1444	805	1280	235	138	207
Effluent COD (mg/L)	320	340	150	76	320	105	81	94
Chemical dose (mg/L)	Fe^{3+}=280	Fe^{3+}=150	Fe^{2+}=840 H_2O_2=1050	Fe^{2+}=1472 H_2O_2=1000	Fe^{2+}=600 H_2O_2=3000	480mg/L/hr	1200mg/L/hr	3000mg/L/hr
Final pH	6.5	7	3.7	3.7	5	7.2		
Contact time (min)	90	85	3.7	140	240	100	30	60
g COD removed/ g Chemical	4.6 w/Fe^{3+}	4.2 w/Fe^{3+}	1.38 w/H_2O_2	0.81 w/H_2O_2	0.43 w/H_2O_2	0.29 w/O_3	0.23 w/O_3	0.07 w/O_3
Sources	Choi E (2007)	Choi Y S *et al.*, (2004)	Choi C W *et al.*, (2005)	Choi Y S *et al.*, (2004)	Park J H *et al.*, (2005)	Choi Y S *et al.*, (2004)		

6.4 APPLICATION OF ADVANCED OXIDATION PROCESSES

Oxidation processes such as wet air oxidation and combustion have been applied to strong liquid and solid wastes. One of these AOP can be combined with coagulation or carbon adsorption to remove residual COD remaining from a biological process. Tables 6.5 and 6.6 give the chemical requirements for polishing biologically treated piggery wastewater effluents and tannin acids, respectively. In practice, combined processes such as chemical coagulation, ozonation, fenton oxidation and carbon adsorption are used for polishing the effluent because of economic reasons. Thus, only the economically applicable processes are discussed as follows:

6.4.1 Fenton Process (Fe^{2+}/H_2O_2)

This process involves the addition of Fe^{2+} and H_2O_2 into a liquid at a low pH. This requires that the final pH must be adjusted following the process and prior to discharge into a water body. The required minimum contact time is 1 hour, but about 2 hours are typically used (Choi *et al.,* 2004; Cho *et al.,* 2005). The optimum weight ratio of Fe^{2+}:H_2O_2 is about 0.8 as shown in Figure 6.7. For a H_2O_2 concentration of 1,050 mg/L, the resulting Fe^{2+} concentration is 840 mg/L. Sludge production is 1.38kg TS/kg Fe^{2+} dose. The weight ratios of COD removed: H_2O_2 required are about 0.8 - 1.3 for piggery waste and 0.82 for tannin acids as shown in Table 6.5 and 6.6, respectively.

Table 6.6 Chemical requirements for removal of tannin acids with different oxidation processes.

Processes	H_2O_2/Fe^{2+}	$H_2O_2/Fe^{2+}/UV$	O_3
COD removed (mg/L): Initial COD=300 mg/L	246	234	176
Chemical dose (mg/L)	Fe^{2+}=276 H_2O_2=300	Fe^{2+}=184 H_2O_2=600	O_3: 1,000 mg/L.hr
Final pH	3	3	3.5
Contact time (min)	5	30	40
g COD removed/ g Chemical	0.82 w/ H_2O_2	0.39 w/ H_2O_2	0.26 w/ O_3
Sources	Lofrano *et al,* 2006		Germirili Babuna *et al* ,2006

6.4.2 Ozonation

The weight ratios of kg COD removed:kg ozone dosed increase as ozone doses are lower as shown in Figure 6.8. This suggests that a lower dose is more desirable. Also, the COD not removed by ozonation can be removed by other processes such as carbon adsorption. The important aspect of ozonation is that the unused O_3 gas must be properly collected and handled. Weight ratios of COD removed:ozone dose are 0.07 to 0.29 as shown in Tables 6.5 and 6.6 for piggery waste and tannins, respectively.

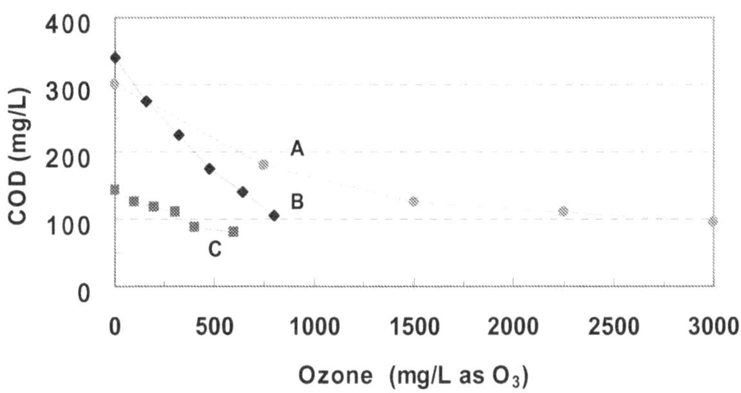

Figure 6.8 COD removal from piggery wastewaters with ozone dose. The ozonation rates are: (A) 3,000 mg/L/hr, (B) 480 mg/L/hr and (C) 1,200 mg/L/hr.

6.4.3 UV Radiation

UV irradiation requires a very shallow flow depth for light penetration which results in a large required surface area (Park *et al.,* 2005).

6.4.4 Electrochemical Destruction

The electrochemical process can be used to remove both nonbiodegradable substances and NH_4-N. Chae *et al.,* (2004) applied this technology to anaerobically digested effluent polished with a microfilter. The electrolyte cell was in a 6 L tank with four (5 × 6 cm) mesh-type plates, DSA (Pt/Nb/Sn-Ti), as the anode and five titanium cathodes with an interelectrode gap of 10 mm. The supply voltage was 7 to 8 V DC and the current varied from 8 to 30A. During the electro oxidation process the solution was continuously recirculated. Factors affecting the electrochemical reactions include current density, electrolyte types and concentrations, anode materials, electrical potential, interelectrode gap and others (Vaze *et al.,* 1997; Polcaro *et al.*, 1998). Chae *et al.,* (2004) study results indicated that COD from digester supernatant and microfilter effluent can be reduced to 100 mg/L within 2 hour contact time. NH_4-N also can be reduced, but a higher efficiency was obtained with the effluent from the microfilter. The addition of an electrolyte improved the NH_4-N removal efficiency. About 20% of the NH_4-N was nitrified to NO_3-N. The amount of electricity consumed for 90% removal of COD and NH_4-N was about 150 kwh/m^3 after microfiltration. This computed electrical requirement was about twice the amount of electricity that can be produced from a digester operation.

6.4.5 Oxidation of Strong Liquids

Oxidation processes applicable to strong liquids include wet oxidation, supercritical wet oxidation and the electron beam process.

(1) Wet Oxidation

This process has been known as the Zimmerman or Zimpro processes. They are operated at high pressure and temperatures from 100 to 372°C. Residence times may range from 15 to 120 min, and COD removal efficiencies may typically be about 75 to 90%. Korea adopted this process in the 1970s to treat nightsoil, which has characteristics similar to piggery wastes, but the plant was later removed because of odor problems, and poor efficiencies resulting in high operating and maintenance costs. In order to reduce energy and material costs, a similar process known as the VerTech system was built underground to a depth of 1,200 m with a diameter of 0.95 m. It uses oxygen instead of air for wet combustion. This system has been used in Apeldoorn, Netherlands since 1994 to treat sludge.

(2) Supercritical Water Oxidation (SCWO)

This process' goal is the complete oxidation of aqueous wastes. Unlike wet air oxidation, which is used to pretreat wastes prior to biological treatment, SCWO is promoted as a single treatment process which can totally destroy the organic pollutants in aqueous wastes. The oxidation reactions in SCWO occur above the critical point of water (374°C and 218 atm) with residence times ranging from 0.2 to 20 min. Organics are converted to CO_2 and H_2O, and nitrogen is converted to NH_3, N_2 and N_2O. However, destruction efficiencies vary with the substances. Acetic acid showed poor destruction, but other organic substances showed more than 90% destruction (Copa, 1997).

(3) Electron Beam

This process uses free radicals (principally the hydroxyl radical, OH^*) to decompose pollutants. In some cases, oxidation via the OH^* is slow and the application of reducing radicals such as the solvated electron (e^-_{aq}) or the hydrogen atom (H^*) is advantageous. An excellent source of both reducing and oxidizing radicals for water treatment is via ionizing radiation produced with the electron beam (Cooper *et al.,* 2005).

Application of irradiation to landfill leachate showed there was a significant improvement (even with its settleability when coagulated) with low irradiation, but it was found that high irradiation was required to oxidized the waste (Electron Beam Accelerator Centre, 2006).

6.5 ADSORPTION

Carbon adsorption has been used for polishing the effluent from AOPs and chemical coagulation. Zeolite can be used for adsorption of NH_4-N. Figures 6.9 and 6.10 show equilibrium concentrations of COD and NH_4-N after contact with adsorbents.

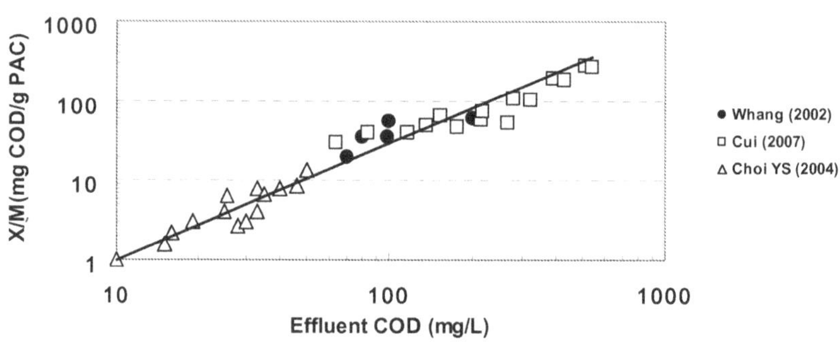

Figure 6.9 COD removal characteristics with carbon. Data from 3 sources are plotted together (Whang , 2002; Cui, 2007; Choi Y S, 2004).

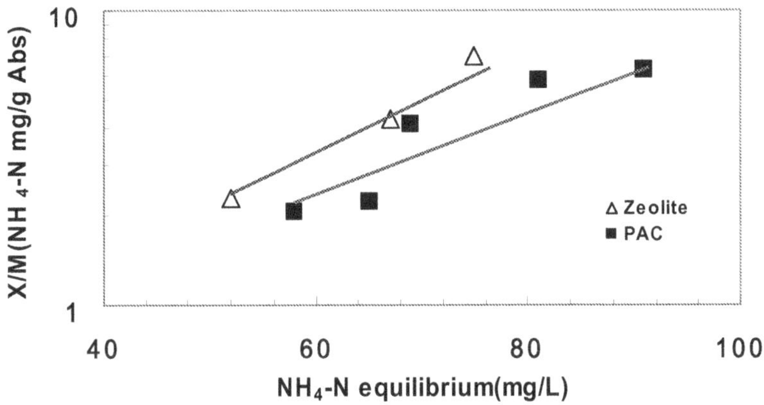

Figure 6.10 Zeolite and carbon adsorption of NH_4-N adsorption characteristics with carbon and zeolite (Cui, 2007).

6.6 MEMBRANE BIOREACTOR (MBR)

MBRs are commonly understood to be the combination of membrane filtration and biological treatment where the membrane primarily serves to replace the clarifier in the wastewater treatment system. The membrane unit can be configured external to, as in sidestream operation, or immersed in the bioreactor (Figure 6.11). The operation of such membrane systems is determined by controlling transmembrane pressure (TMP) and cross-flow velocity, which are generated from a pump. Immersed systems differ in that there is no recirculation loop as the separation occurs within the bioreactor itself (Stephenson *et al.,* 2000). The first generation of MBRs applied the use of cross-flow operated membranes installed in units outside the aeration tank with low velocity circulation pumps. This required a high energy to generate sufficient velocities across the membrane surface. The development of submerged low pressure configuration by immersing the membranes into aeration tank was an important step in making viable commercial solutions for the MBR process. A variety of process configurations exist where the systems are designed to be operated under low pressure or vacuum (Leiknes T and Odegaard H, 2005). Application of membranes to piggery waste is basically for the improvement of effluent produced from biological processes. MF, UF, NF and RO membranes have been applied in Korea for piggery waste treatment system to comply with various effluent water requirements. Table 6.7 and

Figure 6.12 show general aspects of pore sizes, required pressures and application purposes of membranes.

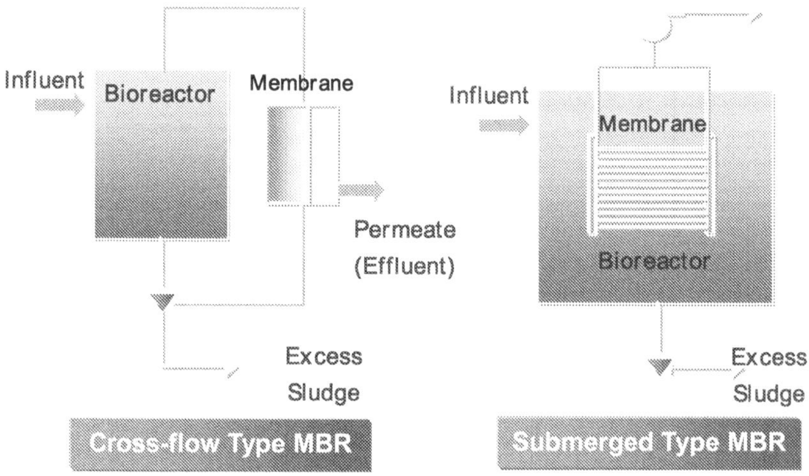

Figure 6.11 Flow types of membrane arrangement.

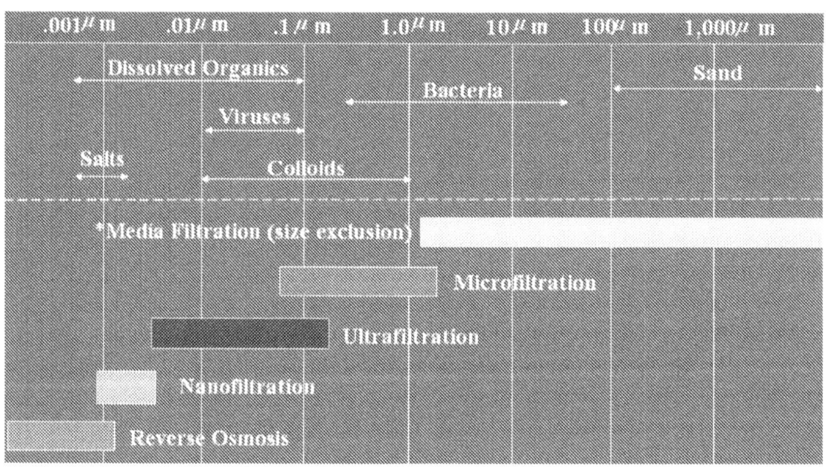

Figure 6.12 Rejection capability of the different membrane separation processes.

Table 6.7 General aspects of membrane configuration, required pressures and applications.

Membrane	Pore size	Pressure	Application
MF	0.1–10μm	<30psi	Removal of suspended matter, turbidity, algae, protozoa, cysts (*Giardia*, *Cryptosporidium*)
UF	1–100 nm	10–100psi	Removal of all of the above + bacteria, viruses and natural organic matter (NOM) (>50,000 Dalton)
NF	200–1000 (MWCO: Dalton)	50–150psi	Removal of NOM and micropollutants Water softening Desalination of brackish water
RO	<200 (MWCO Dalton)	150–600psi	Removal of recalcitrant micropollutants (arsenic) Desalination of brackish and seawater Production of ultrapure water (electronic industry)

Note: psi × 0.0703 = kg/cm^2

Flux rates of MBRs range from 5 to 300 $L/m^2/hr$, with specific flux values ranging from approximately 20 to 200 $L/m^2/hr/bar$ which vary with TMP, cross-flow velocity, pore size and biomass characteristics. Sidestream systems, in general, generate higher flux rates than submerged systems by operating at higher pressure. The greater pressures require greater levels of liquid shear at the membrane surface to control fouling. Operating conditions have varied with pressure of 1 to 5 bar and cross flow velocities around 1 to 3 m/sec (Stephenson *et al.*, 2000).

Table 6.8 Water quality changes through different membranes installed in combined treatment of piggery and kitchen wastewater in PJ, Korea.

Constituents (mg/L)	Effluent qualities through unit processes			
	O_X*	MF	NF	R/O
TCOD	13,900	-	-	-
SCOD	2,900	1,500	270	7
TSS	6,500	-	-	-
VSS	-	-	-	-
TKN	-	-	-	-
NH_4-N	20	18	10	-
NO_2-N	200	180	-	-
NO_3-N	-	-	-	-
TP	-	-	-	-
PO_4-P	62	62	29	-
pH	8.6	8.6	8.6	8.0
AlK	3,400	3,200	1.900	-
Cl^-	4,000	3,800	3,100	107

*Ox denotes oxic unit in membrane bioreactor.

Installed membrane characteristics for 224 m3/d with processing water:

- MF (0.2 μm pore size, made of pp, 1 m^2-900 flat sheet membranes): 14 m^3/hr (operating = 7.8 $L/m^2/hr$), 7kg/cm^2, yearly replacement of 1/3 installation
- NF (Hydranautics, 8 inch diameter spiral wounded polyamide membrane with 40 inch L, 12 elements): feed 13.4 m^3/hr→9.4 m^3/hr (operating =15.8 $L/m^2/hr$), recovery 70%, 10 to 25 kg/cm^2 (operating with 14 kg/cm^2), expected replacement: every year
- RO (Hydranautics, 4 inch diameter spiral wounded polyamide membrane with 40 inch L, 28 elements): operating flux =34 $L/m^2/hr$, replacement every year.

6.6.1 Application of membrane for piggery and kitchen waste treatment plant

Table 6.8 shows water quality changes through different membranes; UF, NF and RO installed for polishing the effluent from biological process treating piggery waste combined with food waste (see the treatment schemes shown in Figure 7.11).

6.6.2 Piggery wastewater treatment plant in CW plant in Korea

Ryu and Kang (2003) present another example of a CW plant with a designed flow rate of 110 to 300 m^3/d for piggery waste. The influent was screened and centrifuged, and the centrate was applied to anaerobic digester. The digester overflow was introduced to anoxic (1,064 m^3) and oxic reactors (3,352 m^3) and the effluent was drum screened (0.75 mm openings) prior to application to UF, which was installed in series as shown in Figure 6.13. During 8 months operation, only water washing was applied to clean UF once every month, and no chemical cleaning was performed. The operating characteristics and results are shown in Table 6.9. The effluent of membrane was designed to be treated by Fenton oxidation followed by activated carbon and sand filtration to comply with the effluent requirement.

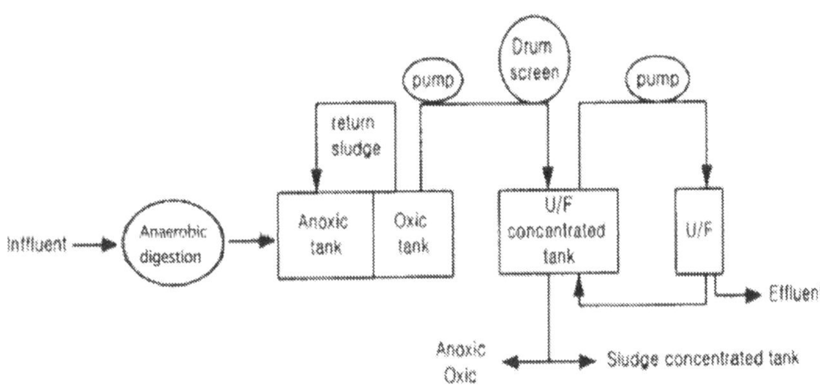

Figure 6.13 Treatment schemes of MBR for piggery waste. Tubular cross flow type membrane with a mean flux of 16.6 $L/m^2/hr$ and pressure of 2 kg/cm^2. The bioreactor operated with 7,000 mg/L MLSS, 3.2 mg/L DO and pH 5.8-6.7 at 33°C.

Table 6.9 Water quality changes through UF in CW plant (mg/L).

	After bioreactor	After UF
BOD	113-642 (205.7)	2.5-25 (9.5)
TSS	770.5	0.2-1.1
NH_4-N		12
TP	18-76.5 (43)	7.4-55 (11.5)
COD		only 20% removal

From Tables 6.8 and 6.9, it is apparent that MF and UF are very successful in removing suspended solids, but do not efficiently remove COD, which basically is the nonbiodegradable fraction as experienced from the tertiary application of municipal effluent. Snyder *et al.,* (2005) stated that the role of membranes and activated carbon in the removal of endocrine disruptors and pharmaceuticals during water treatment are as follows:

- RO is a superior tech, but trace level (ng/L level) contaminants still exist
- UF is not effective, but good for steroid hormone removal
- MBR showed marginal improvement over activated sludge method
- Carbon adsorption is effective. More hydrophobic removal than hydrophilic
- No single process successful, need combination of membrane, carbon + AOP

6.6.3 VSEP RO (vibrating shear extraction process reverse osmosis)

RO is very effective in removing COD, nitrogen and phosphorus. VSEP RO membrane is also used for polishing piggery waste. About 130 m^3/d of piggery wastewater collected from nearby small-sized farms is treated biologically followed by VSEP RO. Table 6.10 is a summary of operating results. Liquid temperature and pH were 27 to 34°C and 6 to 6.5, respectively. Although high shear rate was created on the membrane surface through vibration, some degree of membrane fouling was unavoidable. The transmembrane pressure went up from 200 to 300 psi (13.6 to 20.4 bar) and the recovery ratio dropped to 65% from 80%, while clean-in-place (CIP) was conducted every two weeks. Installation of UF combining RO process, higher recovery was maintained over 3 weeks and CIP period was elongated to more than 2 months. Chi *et al.* (2000) studied different types of VSEP and their results are shown in Table 6.11.

Table 6.10 Treatment efficiencies with VSEP (modified from Lee *et al.*, 2004).

	Influent conc.	Pretreatment		BNR		VSEP RO	
		Conc.	Rem. eff.	Conc.	Rem. eff.	Conc.	Rem. eff.
BOD	4750-8780	3760-6000	22.2	52-690	96.4	1.5-12	97.6
CODcr	9000-27700	5600-25200	30.2	570-3170	89.4	12.5-145	96.6
SS	1050-37800	1100-12400	41.1	63-1230	91.4	0-2.6	99.7
TN	3390-7300	2240-5570	32.1	137-1040	88.8	8.6-28	96.2
TP	96-261	61.4-178	25.2	37.2-192	51.5	0.8-3.8	96.7

Note) Conc in mg/L and rem eff in percent.

Table 6.11 Water quality changes through different types of VSEP (Chi *et al.*, 2000).

Constituents	Raw waste water	ACM	ESPA	BW-30
COD	185	3.6	0.6	2.6
SS	22.6	0.5	0.3	0.9
TN	116.08	11.75	3.75	4.29
TP	52.2	0.07	0.03	0.04
TDS	1300	38	43	28.4
Conductivity (μs/cm)	1987	51.3	63	47
Na^+	113	4.3	2.98	4.11
NH_4^+	79.55	1.20	1.50	0.89
K^+	434.4	8.5	7.5	7.5
Mg^{2+}	24.6	0.2	0.13	0.12
Ca^{2+}	41.14	3.11	2.50	2.19
F^-	0.25	ND	ND	ND
Cl^-	146.43	8.5	7.43	5.10
NO_3^-	2.39	ND	ND	ND
PO_4^{3-}	68.3	1.20	0.4	0.13
SO_4^{2-}	188.13	5.9	4.4	4.65

Units: mg/L except where noted.

One of the problems in a RO process is the disposal of the byproduct concentrate. The concentrate of the VSEP RO process is an odorless dark brown liquid with 1,650 mg/L N, 382 mg/L P and 2,059 mg/L K. The color is caused by dissolved humic acid. The chemical composition of the concentrate is shown in Table 6.12. It was found that the concentrate can be used as a low concentration liquid fertilizer (Yoon *et al.,* 2004). The yield of rice (about 6 tons of hull/ha at 110 kg N/ha/yr) is similar to that of chemical fertilizer applied. The water extractable nitrate content of the soil dose did not exceed that which resulted when a chemical fertilizer was applied at soil depth of 30 and 60 cm. The amount of nitrate which percolated into the water table also showed a similar level with chemical fertilizers. There was no significant deposition of phosphorus at a rate of 110kg N/ha, but a significant increase of labile phosphorus was observed at a rate of 220 kgN/ha with the concentrate.

Table 6.12 Chemical composition of the concentrates (modified from Yoon *et al.,* 2004).

Constituents	Concentration (mg/L)	Constituents	Concentration
N	1,650	Org – Carbon (%)	45.5
P	382	Total acidity	5.17
K	2,059	COOH group ($C_{mol(+)}/g$)	2.38
Ca	191	Phenolic OH group	2.79
Mg	143		
Na	125		
Zn	9.9		
Cu	1.8		
Fe	8.7		
Mo	0.01		
Mn	1.0		

REFERENCES

Bowers A R (1997) *Fundamentals of chemical oxidation in principles and practices of chemical oxidation in wastewater treatment*, Vanderbilt University, Double Tree Hotel, Nashville, TN.

Chae K J, Yim S K, Choi K H, Kim S K and Park W K (2004) Integrating biological end electro-chemical treatment of swine manure, *Wat. Sci. Technol.* **49**(5-6), 427-414.

Chi E-S, Kim J-W, Shin D-Y (2000) The removal of nitrogen and phosphorus for the swine wastewater by VSEP membrane system, *Korean J. Sanitation,* **15**(3), 31-38.

Chinese regulation (2001) GB18596-2001, ROC

Cho C W, Chae S C, Soo C, Park S K, Youn J, Chung P G (2005) Degradation characteristics of Non-biodegradable matters using pre-coagulation and Fenton oxidation process in livestock wastewater, *Annual meeting of Korean society of water quality and Korean society of water and wastewater.*

Choi E (2007) *Integrated technology development of animal waste treatment and recycling, final report,* Korea Institute of Environmental Science and Technology. Seoul, Korea.

Choi Y S, Hong S W, Kwon G H and Jung I H (2004) Comparisons of physical and chemical methods for dealing with biologically pre-treated livestock wastewater as a post-treatment, *J KSWQ*, **20**(2),110-119.

Copa W M (1997) Chemical oxidation symposium workshop hydrothermal processes, US filter, 715-359-7211, *Principles and practices of chemical oxidation in wastewater treatment, Vanderbilt University*, Nashville, Tenn.

Cooper W J, Gehringer P, Pikaev A K, Kurucz C N and Mincher B J (2005) Radiation processes in *Advanced oxidation processes for water and wastewater treatment* (edited by Simon Parsons, IWA publishing).

Cui M C (20007) *Evaluation of highly loaded piggery wastewater treatment system: a case study*, MS thesis, Department of civil and environmental engineering, Korea University, Seoul, Korea.

Elecron Beam Accelerator Centre (2006) Yeungnam University, Gyeongsan, Korea (www.ebac.re.kr).

Germirli Babuna F, Yilmaz Z, Okay O, Alaton Arslan I, Iskender G (2006) Ozonation of synthetic versus natural textile tannins: Recalcitrance and toxicity towards Phaeodactylum tricornutum, *7th International symposium on waste management problems in agro-industries Amsterdam,* 27-29.

Hwang I S and Min K S (2006) Piggery waste treatment using improved MLE process full scale, *JKSWQ*, **22**(5), 895-904.

IARC(International Agency for Research on Cancer, WHO) (1976) Monographs on the evaluation of carcinogenic risk of chemicals to man, some naturally occurring substance, CAS No: 1401-55-4, volume 10, *Tannic Acid and Tannins* , 253-262.

Jacangelo, J G (1995) Selected processes for removing NOM: An overview, AWWA, 87(1), 64

Lee J-Y, Choi E, Park Y K, Lee H-W (2000s) Electrochemically enhanced activated denitrifying biofilm process, *Annual meeting of KSWQ and KSWW*, Seoul, Korea (Korean).

Lee J, Kim S, Choi J, Hwang H, Cha M and Kim J (2004) Tertiary treatment of biologically treated piggery wastewater using vibratory shear-enhanced RO membrane, *Wat. Sci. Technol.* **49**(5-6), 435-442.

Leiknes T and Odegaard H (2005) The development of a biofilm membrane reactor, *WRRS2005*, Cheju, Korea.

Lofrano G, Meric S, Belgiorno V, Nikolaou A, Napoli RMA (2006) Fenton and photo-Fenton treatment of a synthetic tannin used in leather tannery: A multi-approach study, *7th International symposium on waste management problems in agro-industries Amsterdam* 27-29.

Polcaro A M, Palmas S, Renoldi F, and Mascia M (1999) On the performance of Ti/SnO2 and Ti/PbO2 anodes in electrochemical method, *J of Korean inst of chem engrs*, **39**(5), 661-668.

Park J H, Chang S W, Cho I H (2005) Optimal condition of operation parameter for livestock wastewater treatment using Photo-Fenton process, *J of KSWQ* 21(3), 284-288.

Park J H, Lee J Y, Kang S H, Lim S H, Moon T H, Anh Y H and Ko K-B (2006s) The effect of nitrate on the removal of 2,4-Dichchlorophenol in the advanced oxidation process using UV and H_2O_2, *Annual Joint Conf of Korean society on water quality and Korean society of water supply and wastewater.*

Park J H, Moon T H, Ahn Y H, Ko K B (2006f) Effect of nitrate on diethyl phathalate degradation in the advanced oxidation process using UV and H_2O_2, *Annual Joint Conf of Korean society on water quality and Korean society of water supply and wastewater.*

Parsons S. (2004) Introduction in *Advanced oxidation processes for water and wastewater treatment,* IWA Publishing.

Ryu J H, and Kang S H (2003) Livestock treatment using membrane bioreactor, *J of Korean society of water quality*, **19**(3), 301-308.

Stephenson T, Judd S, Jefferson B and Bindle K (2000) *Membrane bioreactors for wastewater treatment,* IWA Publishing.

Snyder S A, Wert E, Yoon Y and Westerhoff P (2005) Role of membrane and activated carbon in the removal of endocrine disruptors and pharmaceuticals during water treatment processes, *WRRS2005*, Cheju, Korea.

Vaze A S, Sawant S B and Pangarkar V G (1998) Electrochemical oxidation of p-t-butyltoluene to p-t-butylbenzaldhyde, *J appl electrochem,* 28, 623-630.

Whang GD, Kang JS, Jang JG (2002) Removal of residual organic matter from the biologically treated swine wastewater by MBR process with PAC addition,. *J of KSEE* **24**(1) 1-10.

Yoon Y, Ok Y-S, Kim D-Y and Kim J G (2004) Agricultural recycling of the by-product concentrate of livestock wastewater treatment plant processed with VSEP RO and bio-ceramic SBR, *Wat. Sci. Technol.* **49**(5-6), 405-412.

7

Integrated Systems

7.1 INTRODUCTION

Piggery waste management can be very simple and very complicated depending on how the waste is used and disposed. When the waste is allowed to be directly applied to land, its management will be very simple. However, it can be very complicated when the waste must be treated and the treated effluent is discharged into a water body, or it is reused as a water supply for animals.

This chapter discusses how piggery wastewater management systems are integrated depending on regulatory requirements. European practice focuses on biogas generation and the total return of the effluent to agricultural land. The US approach is focused on Innovative Sustainable Systems Utilizing Economical Solutions (ISSUES) with "Environmentally Superior Technologies (ESTs) using existing lagoon and sprayfield technologies. Unlike Europe and the US, Korea allows treated effluent to be discharged into water bodies. There are many centralized plants which operate on this basis and whose effluents comply with the effluent regulations as shown in Table 1.2.

7.2 MANURE BASED BIOGAS PLANTS (MBBP)

Low tech anaerobic digestion (AD) with energy recovery followed by land application is the common practice in Europe and the US. However, unlike the US, in Europe there are many MBBP (manure-based biogas plants) which use numerous types of inputs beside manure These include: farm products, grass, alfalfa, straw, cabbage, potatoes and other organic wastes including sewage sludges. The recovered biogas is mostly used for the production of combined heat and power (CHP). The biogas can be used as a source of fuel for gas heating systems, boilers, fuel cells, microturbines, vehicles and other uses. Since the prices of coal, oil and natural gas are rising, these biogas systems will become even more attractive. As the unit size increases, they are more economical. Biogas plants are typically built in Europe in the range of 300 to 2,000 KWe (KW in electricity). A 500 KWe facility is able to produce energy for about 0.08 to 0.09 EUR/KWe (German prices). With investment costs from 2,500 to 3,000 Euro/KWe and reactor volumes ranging from 1,000 to 10,000 m^3, the centralized MBBP are comparable in costs to small wastewater treatment plants designed for 1,000 to 10,000 population equivalents. The average electricity production capacity of approximately 2,000 MBBP in Germany is only 125 KWe. A two-stage, mesophilic MBBP with a TSS feed of 8 to-10% solids, an electricity production capacity of 500 KWe or a thermal production capacity of 700 KWt (KW in thermal energy) cost about 1.3 million euro. The feedrate was 46 ton/d of cattle and pork liquid manure along with the use of energy crops like maize, rye and straw. This plant is typical to many MBBP in Germany (Wiese and Haeck, 2006).

Although AD is a well known technology for the treatment of strong wastes, its use by farms has not been common. However, interest in AD technology is growing because of environmental reasons and also because it is a means of renewable energy production from mixtures of energy crops, agricultural residues and commercial food wastes. Banks *et al* (2006) summary on the use of the AD process in the EU follows:

> Denmark was the first European country to encourage and coordinate the widespread use of AD for agricultural wastes. In 1988, a Biogas Action Programme was established with R&D goals along with the collection of information from the construction and monitoring of AD plants. Also, it included economic measures to encourage investments in centralized plants, providing grants of up to 40% of capital costs with long-term and low-interest loans, tax exemptions and subsidies for bioelectricity generation. Further, there was an incentive for heat sales from combined heat and power (CHP) plants for 6 to 9 months per year through district heating networks. The program is an example of a coherent and ambitious government policy that

was aimed initially at improving farm management and nutrient control practices.

Areas used for farming in Denmark are designated as vulnerable under the EU Nitrate Directive. There is a requirement for 9 months of manure storage plus restrictions on the application and landfilling of organics. As a result of economic subsidies and the need to comply with environmental directives, there are now 20 centralized and over 35 farm-scale plants in Denmark providing treatment to over 1 million tones of animal wastes per year. However, due to policy changes, and in particular the introduction of the bill to amend the Act on Subsidies for Electricity Production," no new centralized plants have been established since 1998 and the development of farm-scale plants has slowed down. The biggest factor in this slowdown is that there is no longer a guaranteed price for electricity generated by new plants.

In comparison, the whole of Finland was declared a nitrate sensitive zone with a requirement for 12 months of manure storage. The government offered no assistance or incentives to farmers for the construction of AD plants. Until recently, there were no grants and no subsidized prices for biogas-generated "green electricity." For example, until 2004, the use of biogas as an automotive fuel was taxed at the same rate as fossil fuel. As a result, interest was limited to enterprising individual cases which indicated that manure management alone was not a strong enough driving force to stimulate the introduction of AD technology.

In Germany, the last 5 years have seen an unprecedented introduction of biogas plants to farms. More than 90% of the 2,500 plus plants thus far in operation use codigestion of animal manure and energy crops for more efficient gas production as shown in Table 7.1. Maize and grass are the most common additions, because maize has a high methane yield per hectare and grass is characterized by low production costs. The motivating factors in Germany are financial incentives and subsidies for renewable energy production (Table 7.2). However, probably the most important factor is the guaranteed price for electricity over a 20 year period. At present, the demand for biogas plants in Germany is outstripping the capacity to construct them.

Table .7.1 Typical biogas yields from manures, industrial/commercial wastes and energy crops (Kottner, 2004)

Manures and slurries	Biogas yield (m^3 $tonne^{-1}$ wet wt.)	Industrial and commercial wastes and residues	Biogas yield (m^3 $tonne^{-1}$ wet wt.)	Agricultural energy crops	Biogas yield (m^3 $tonne^{-1}$ wet wt.)
		Potato distillery	35	Meadow grass	98
Dairy cattle	20	Vegetable processing	35	Maize silage	190
Fattening cattle	34	Rape seed cake	612	Grass silage	183
Pig	18	Canteen residues (high fat)	90	Milled grain	597
Poultry	93	Canteen residues (low fat)	44	Corn crop mix (5.3% fibre)	391
		Flotation fat	108	Total plant grain silage	195

Table 7.2 Basic electricity price and bonuses paid for farm-based digestion in Germany

Payment/Ouput capacity	< 150 kW_e	< 500 kW_e	< 5 MW_e	> 5 MW_e
Basic electricity price	11.5	9.9	8.9	8.4
Bonus for energy crops	6	6	4	-
Bonus for CHP	2	2	2	2
Technology bonus (only if CHP condition fulfilled)	2	2	2	2

In the UK, there has been little development of farm-based AD despite the relatively high price for "green electricity" provided by the Renewable Obligations Certificates (ROCs) scheme. The one plant built to date, with an EU grant, follows the Danish model of centralized digestion. In this scheme, slurry from 25 farms is handled with imported food waste in order to boost energy production. Since coming on stream in 2002, the scheme has been plagued with regulatory problems, mainly because of the need to comply with the EU Waste Framework Directive. This is a result of the interpretation that once animal manure leaves the farm (as is the case in any centralized scheme), it becomes a waste which is subject to licensing requirements for transport, handling, processing and reapplication to the land. The financial penalties of such a regulation, and the perception by farmers that manure, which can be spread on land untreated and is not regarded as a waste, is "transformed into a waste," under the regulation has been a major obstacle to the replication of the concept elsewhere. Perhaps the biggest disincentive is that there is no guaranteed price for energy, making the commercial case for banking support difficult.

The potential of biogas has not yet been realized, as it is mainly produced locally, far from the major energy consumers found in urban and industrial areas. Therefore there is a tendency towards centralization of plants with very large scale facilities to reduce costs and increase earnings. The large amount of biogas can be injected directly into the natural gas grid, as is already done in Sweden and Austria. Alternatively, methane (upgraded biogas) can be used as a car fuel or in combustion engines for local electricity production. Also, low cost and environmentally friendly electricity production technologies are available on site. A number of newly developed technologies, such as fuel cells and microturbines, are used to convert biogas into electricity with a greater efficiency than current technologies. Fuel cells are highly efficient because they transform chemical energy into electricity. Electricity can be transported very easily from the point of production to the consumption location. Fuel cells are easily scaleable and small scale energy production can be used for electric generation. For example, the HXS 1,000 Premier fuel cell system has a nominal power rating of one KWe of electricity and 3 KWt of heat. Consequently, the electric efficiency is about 33%. Sulzer demonstrated this use of biofuel by installing an HXS Premier on a farm in Switzerland (Van herle *et al.,* 2004; Lens, 2006).

It is theoretically attractive to operate anaerobic digester at thermophilic conditions, which produces gas with a shorter retention and a sludge with less pathogens. However, thermophilic anaerobic digestion of swine manure yields low quality methane because ammonia and sulfide gases are produced. Also, a higher concentration of volatile fatty acids are produced and accumulated.

Several methods have been tested in order to increase the methane yield. Activated carbon or iron chloride additions have increased methane yields (Hansen *et al.,* 1999). When an energy recovery system with combined heat and power (CHP) is utilized, the excess heat from electrical generation can be used for heating the digesters. If there are two stage digesters, the first digester can be operated at mesophilic conditions to produce methane gas and the second digester can be heated to thermophilic conditions to kill pathogens, but with little gas production. This combined operation may satisfy both energy production needs as well as accomplishing pathogen control without any loss of energy.

The economical size of a commercially available internal combustion engine generator, which can be fueled with methane gas, is some where between 100 to 150 KWe. Smaller digesters may require the use of microturbine technology. Microturbine and small internal combustion engine generator technologies are compared as shown in Table 7.3.

Table 7.3 Comparison of the operating characteristics of electric generation systems and microturbine technology for combined heat and power (Wiltsee and Emerson, 2003).

Technologies	IC*engine generator	Microturbine
Status	Proven	Relatively new, unproven
0il changes and overhauls	Frequent	Infrequent
Moving parts	Large number	Small number
Tendency to break down	High	Potentially low
Maintenance	High	Unmanned, remote control operation
Operation	Noisy and dirty	Quiet and clean
Emissions of NOx, CO and other pollutants	High	Extremely low
Sizes available (KWe)	50-800	30-250
Operating pressure		6 atm
Input status		Dry
Installation	Commonly used	LA Sanitation District

*IC = internal combustion

Lessons learned in early digester gas microturbine projects are summarized by Wiltsee and Emerson (2003). When clean, dry fuel is supplied reliably to microturbines, the systems operate continuously without any attention required for months at a time. However, drying of the compressed digester gas is very important since some compressor failures have been caused by the condensate. Compounds in the condensate foul the microturbine fuel control valves and fuel injectors. Refrigerated dryers have proven to be reliable for the removal of condensate. Gas/liquid separation and the design of the compressor cooling system are important. Materials of construction must address the corrosive properties of the digester gas. Stainless steel is best. Carbon steel and yellow metals are not recommended.

7.3 ENVIRONMENTALLY SUPERIOR TECHNOLOGIES (EST)

One of the most extensive projects to develop and implement advanced swine waste management systems was conducted in North Carolina, USA. In the Summer and Fall of 2002, the Attorney General of North Carolina entered into agreements with Smithfield Foods, the nations largest pork producer in the US, and its subsidiaries and then with Premium Standard Farms under which the two companies consented to fund the development of EST for use on North Carolina swine farms owned by the companies. The 18 candidate technologies, including solids separation, solid and liquid management, byproduct recovery and tertiary treatment for reuse, are shown in Figure 7.1. The 18 candidate technologies identified for development were based on the following five criteria (Humenik *et al.,* 2004) and were finally developed as Innovative Sustainable Systems Utilizing Economical Solutions (ISSUES) (Bull, 2005):

(1) Eliminate the discharge of animal waste to surface waters and groundwater through direct discharge, seepage, or runoff.
(2) Substantially eliminate atmospheric emissions of ammonia.
(3) Substantially eliminate the emission of odor that is detectable beyond the boundaries of the parcel or tract of land on which the swine farm is located.
(4) Substantially eliminate the release of disease-transmitting vectors and airborne pathogens.
(5) Substantially eliminate nutrient and heavy metal contamination of soil and groundwater.

The goal of "ISSUES as EST" is to utilize, incorporate and maximize the existing infrastructure, and provide EST that would become both operationally and economically feasible for the swine industry. ISSUES consist of three separate approaches to swine management, which build upon existing lagoon and sprayfield technologies. The three approaches are:

Aerobic Blanket Technology (ABS)
Permeable Cover System (PCS)
Recycling of Existing Nutrients, Energy and Water (RENEW)

The three systems are introduced in more detail to show how these systems work. In addition, some other important processes, not explained earlier, are discussed.

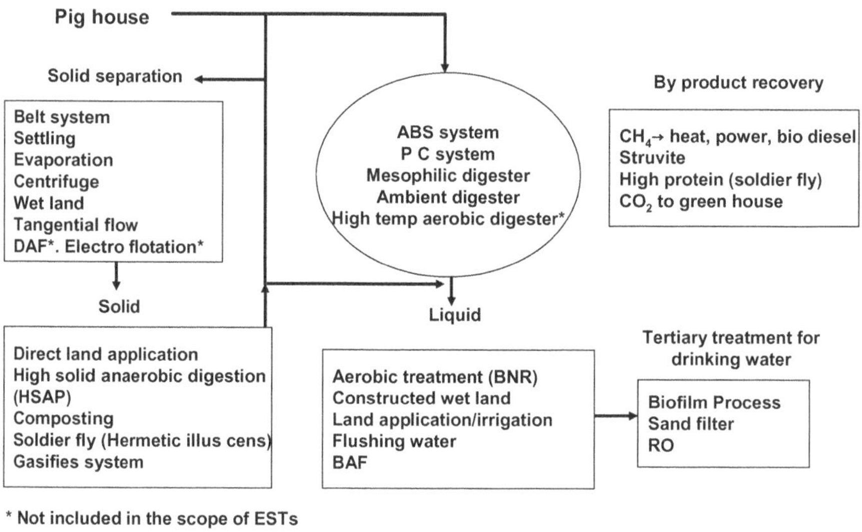

Figure 7.1 Scope of EST for animal waste management.

7.3.1 Aerobic Blanket System (ABS) Technology

The basic goal of the ABS was to reduce odor and ammonia emission from anaerobic lagoons. Under steady-state conditions, the total system resulted in a reduction of nitrogen fractions of 27 to 33% with an increase of 59% in NO_3-N. It was assumed that the reductions were a result of reduced ammonia and/or denitrification emissions and the retention of organic forms associated with sludge biomass accumulations. A 40% solids reduction was observed across the complete system. The retention of phosphorus, zinc and copper occurred at rates of 50% P, 64% Zn and 35% Cu, respectively. These retentions were assumed to be part of the biomass and sludge accumulations.

Figure 7.2 shows a schematic of the ABS system. Pig houses discharge wastewaters (50,000 gpd or 200 m^3/d) consisting of feces, urine and flushing water into an anaerobic lagoon (12.1 MG or 48,000 m^3). The anaerobic lagoon effluent was pumped to an aerobic digester or an aerobic lagoon with a capacity of 5MG (20,000 m^3). The nitrified aerobic digester effluent was pressurized with air and recycle back into the surface of the anaerobic lagoon as mist through nozzles located above the anaerobic lagoon. The air helped to nucleate the water which formed adsorption centers for potentially volatile compounds. A 7.5 hp compressor was used to pressurize the air which was stored in a 660 gallon (2,500 L) tank at 125 psig (8.8 kg/cm^2). A 1.5 hp pump was used to feed nitrified water into 24 nozzles at 25 gpm (94.6 Lpm). The liquid was sprayed above the surface of the anaerobic lagoon every 15 minutes for approximately 1.27 minutes. This delivered approximately 2,400 gallons (9,000 L) of mist per day. No spraying occurred if the wind speed was greater than 10 mph (16 km/hr). This system was applied at a farm with 6,480 finishing head (720 head per house). Each house was flush style and was mechanically ventilated.

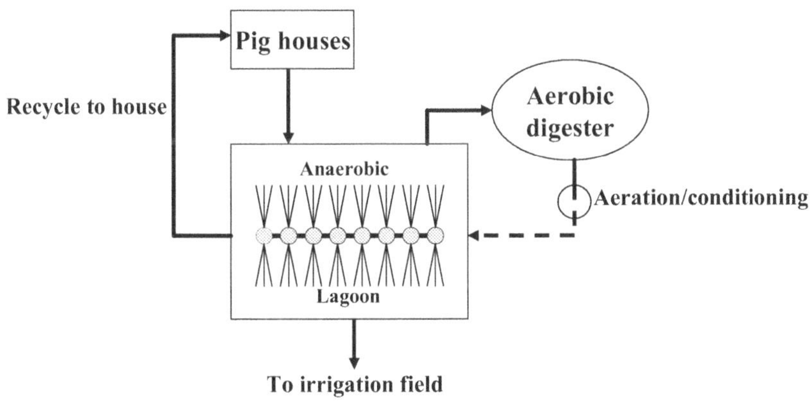

Figure 7.2 A schematic flow diagram of an aerobic blanket system (ABS) used to reduce odor and NH_3 emissions form lagoon (modified from Bull, 2005).

7.3.2 Permeable Cover System (PCS) Technology

An application of the PCS system included an anaerobic treatment lagoon with a permeable cover made of woven polypropylene (BioCap^{ML}) to facilitate the management of nitrogen and nutrients, and to reduce emissions of ammonia and odor (Miner *et al.,* 2003). An aerobic digester was used to enhance nitrification. The effluent from the aerobic digester was used to clean the pig house, and excess liquid was stored for subsequent land application and spray onto the permeable cover for to promote evaporation. A schematic flow diagram of the installed PCS plant is given in Figure 7.3.

Use of the PCS technology resulted in a 51% reduction of total N across the entire system and 46% reduction for NH_4-N. As observed in the ABS system, a seasonal effect (temperature) on NH_4-N was recorded in the aerobic digester with lower values in warmer seasons. The permeable cover was designed to act both as a barrier and as a biofilter (or matrix for aerobic bacteria growth) used to reduce lagoon emissions.

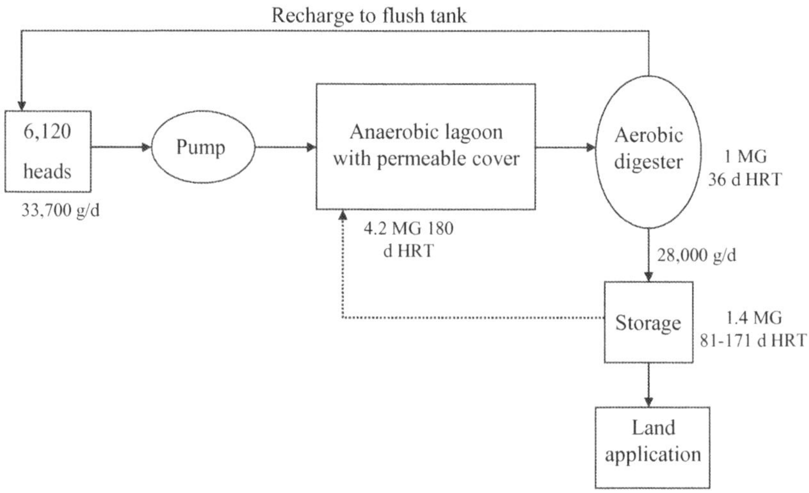

Figure 7.3 Schematic flow diagram of an existing lagoon system with a permeable cover (modified from modified from Bull, 2005). G denotes gallon which is equivalent to 3.785 L. MG =million gallons

7.3.3 Recycling of Existing Nutrient, Energy and Water (RENEW)

The applied RENEW process included equalization, a thickener, anaerobic digester (mesophilic), storage and water treatment as shown in Figure 7.4. It was designed to handle the waste discharged from eight pig houses (1,224 head/house or 9,792 head) estimated to be 77,000 gpd (2.3 gallons of wastewater/pig/d or 8.7 L/head/d). Only thickened sludge (estimated to be 34,000 gpd or 130 m^3/d) was to be fed to the digester, but the actual operation involved the addition of unthickened sludge and influent into the anaerobic digester. The anaerobic digester effluent was stored prior to application to the aerobic digester for nitrification. Its effluent was designed to be used to recharge each house's flush tank or to be reused as a source of drinking water for the animals after further water treatment, which included a clarifier, a sand filter, reverse osmosis, equalization and a UV filter. The anaerobic digester produced 35 SCF gas/kg VS removed (0.98 m^3 gas/kg VS removed) with 65 to 70% CH_4, 30 to 35% CO_2 and 0.05% H_2S and NH_3. The VS removal efficiency of the digester was estimated to be about 71%. The peak gas yield was 46,000 SCFD (1,300 m^3/d) which is equivalent to 30 MBtu/d (7.56 M Kcal/d). The heating value based on volume was 5,900 kcal/m^3 gas. Each pig produced 0.13 m^3 gas/d.

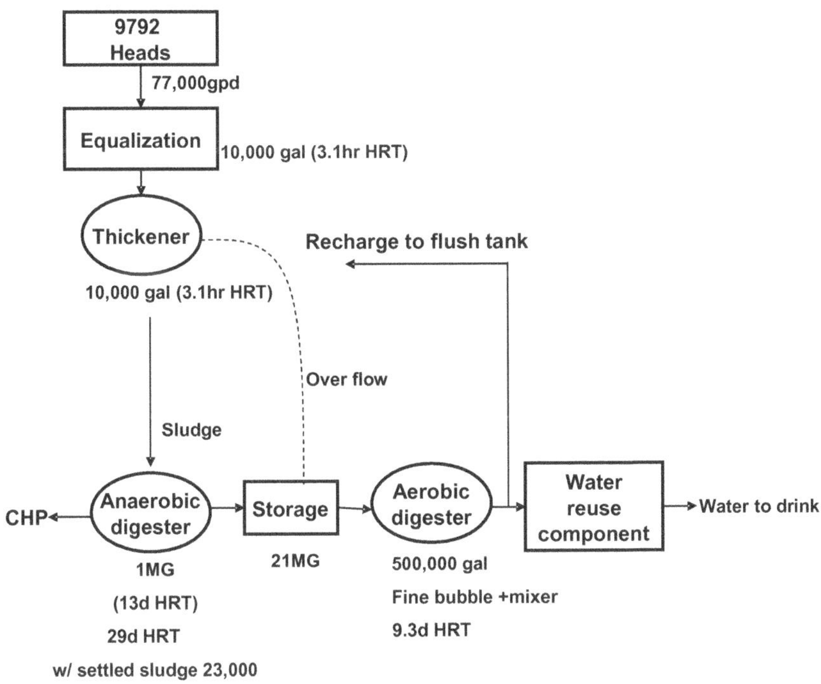

Figure 7.4 Schematic of the RENEW process (Modified from Bull, 2005)

7.3.4 High Solid Anaerobic Digestion (HSAD)

HSAD is an innovative anaerobic technology based on film transfer theory involving digestive products and anaerobic microbes in a solid environment. HSAD was applied to army food waste (Bull and Cook, 2004). It utilized a sub-rpm, low shear and pug-tine agitation to protect and preserve the bacterial films required for efficient digestion and to maintain the constant contact between the hydrolytic bacteria and substrate necessary for rapid breakdown and solubilization of polymers, and a bacterial consortium acclimated to ammonia concentrations that exceeded 4,000 ppm. The total N was reduced by approximately 50% during HSAD processing. Methane gas was produced up to 2.7 m^3/m^3 feed. Other design and performance aspects of the process included:

Feed solid contents > 30%
VS loading = 5 $kg/m^3/d$
Influent COD = 500 g/m^3
Operating temperature = 55 to 60°C
330 L CH_4 production/kg COD removed with a 55% CH_4 content in the gas
Pathogen kill > 99%

7.3.5 Biological Aerated Filter (BAF) System

The BAF system used for piggery waste treatment included a solid/liquid separation process with liquid equalization followed by two BAF stages as shown in Figure 7.5. This system was used for five pig houses with 4,300 finishing pigs. The flow from the equalization tank was 500 to 1,000 m^3/d flow with 2,000 to 4,000 mg/L COD, 270 to 750 mg/L BOD, 100 to 500 mg/L NH_4-N and 1,800 to 2,660 mg/L SS. Each BAF unit was 151 m^3 filled with 130 m^3 of plastic media. The average COD load was 6 kg COD/m^3 biofilter media. The liquid temperature greatly affected BAF process performances. COD, BOD and nitrogen removal efficiencies were greater than 75% at 27°C, but the efficiencies were lowered to 50 to 70% at 10°C. Mass balances indicated that COD and TN removal efficiencies were 37 to 67% and 46 to 69%, respectively. High concentrations of NO_x-N remained in the effluent. The BAFs were backwashed to prevent clogging.

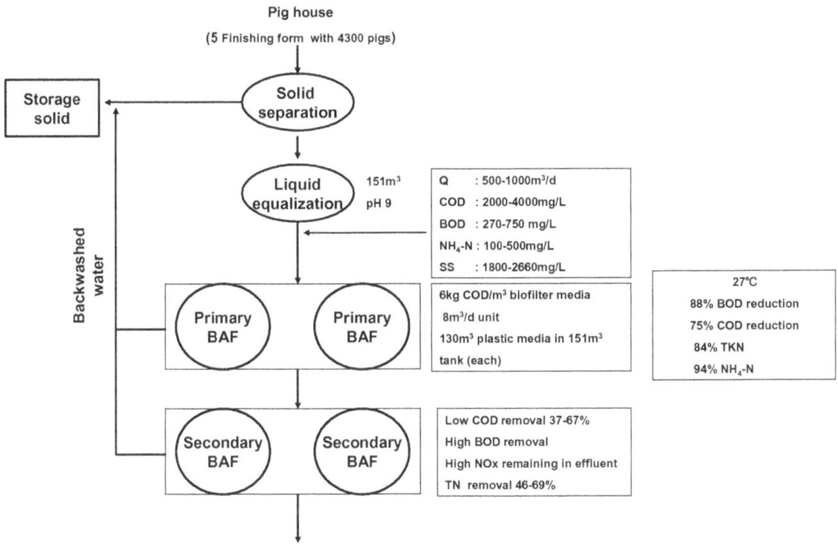

Figure 7.5 Schematic flow diagram of a biological aerated filter (BAF) system to piggery waste (modified from Bull, 2005).

7.3.6 BEST Biofuel System

Smith Foods, Inc. installed a biofuel system known as Biomass Energy Sustainable Technology (BEST) with an anaerobic digester. The system is for 23 operations totaling 257,000 finishing hogs at the Skyline complex of the Circle Four Farms in Utah. The daily flow rate is 6,500 m^3/d, which is equivalent to 38,000 m^3/d of sewage a day. Figure 7.6 is a schematic flow diagram of the biodiesel production system described by Dugba (2003).

The system includes three buffer basins and four digesters. The wastes are gravitationally thickened and fed to digester with 4.5% solids. The anaerobic digesters were constructed in-ground and completely enclosed in multiple liners of a high density polyethylene (HDPE) geomembrane. A total of 35,000 m^3 gas was produced daily with 65 to 70% CH_4 at 35°C. This gas was collected and fed to the biomethanol plant, of which the key design components include: (1) gas scrubbing tower, (2) pressure vessels packed with proprietary catalysts that serve as a thermochemical reactors where the steam reformation and biomethanol formation takes place, (3) distillation towers to purify and separate the biomethanol from other products and reactants and (4) cooling towers to dissipate excess heat. The BEST process is adopted from a process developed in

1923 by Franz Fischer and Hans Tropsch, also known as the Fisher-Tropsch (FT) process. Unlike the biogasification process, the FT process is nonbiological and therefore is a practicable and mature technology. The biogas produced in the anaerobic digestion of swine manure is already at the desired stoichiometry (~65:35 of CH_4:CO_2) and therefore there is no need to purchase CO_2 for the production of synthesis gas (a mixture of carbon monoxide and hydrogen) as shown in Table 7.4.

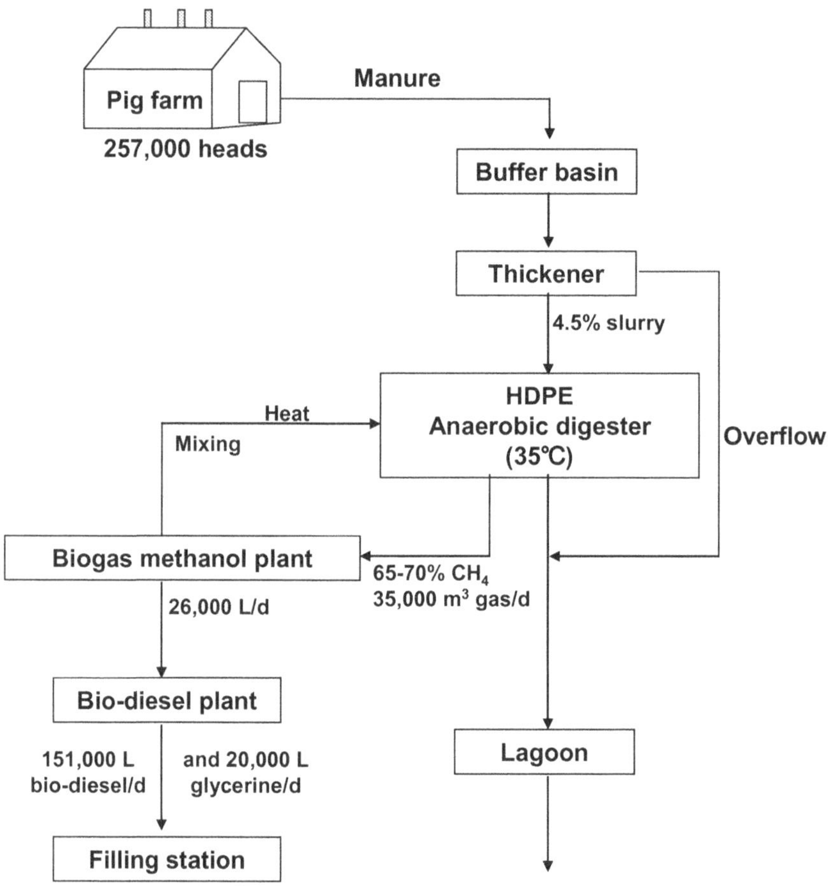

Figure 7.6 Schematic flow diagram of the Smithfield BEST biofuel system (modified from Dugba (2003)).

Table 7.4 Chemistry for converting pig manure to biodiesel as presented by Dugba (2003).

Hydrolysis of manure:

$C_6H_{13}NO_5$ (manure) + $H_2O + H^+ \rightarrow C_6H_{12}O_6 + NH_4$ ....(7.1)

Acetogenesis:

$C_6H_{12}O_6 + 0.1115NH_4^+ \rightarrow 0.1115\ C_5H_7NO_2 + 0.744\ CH_3COOH +$
$0.5\ CH_3CH_2COOH + 0.5CH_3CH_2CH_2COOH + 0.454CO_2 + 0.446H^+$..................(7.2)

Propionate Hydrogenesis:

$CH_3CH_2COOH + 1.786H_2O + 0.0458NH_4^+ \rightarrow 0.0458\ C_5H_7NO_2 +$
$0.924CH_3COOH + 2.778H_2 + 0.1832H^+ + 0.0924CO_2$ .... (7.3)

Butyrate Hydrogenesis:

$CH_3CH_2CH_2COOH + 1.84H_2O + 0.0545NH_4^+ \rightarrow 0.0545\ C_5H_7NO_2 +$
$1.86CH_3COOH + 1.92H_2 + 0.2180H^+$.. (7.4)

Homoacetogenesis:

$2.073H_2 + CO_2 + 0.0487NH_4^+ \rightarrow 0.0487\ C_5H_7NO_2 + 0.378CH_3COOH +$
$1.146H_2O + 0.1948H^+$ .... (7.5)

Hydrogen Methanogenesis:

$3.813H_2 + CO_2 + 0.022NH_4^+ \rightarrow 0.022\ C_5H_7NO_2 + 0.89CH_4 +$
$1.956H_2O + 0882H^+$..(7.6)

Acetate Methanogenesis:

$CH_3COOH + 0.022NH_4^+ \rightarrow 0.022\ C_5H_7NO_2 + 0.945CH_4 +$
$0.945CO_2 + 0.06H_2O + 0.0882H^+$ .... (7.7)

Biogas to Biomethanol:

Equations 8 and 9 are the key chemical reactions that take place in the steam/methane reforming process.

$CH_4 + H_2O \leftrightarrow CO + 3H_2$ ....(7.8)
$CO + H_2O \leftrightarrow CO_2 + H_2$ ....(7.9)
$CO + 2H_2 \leftrightarrow CH_3OH$... (7.10)

Transesterification

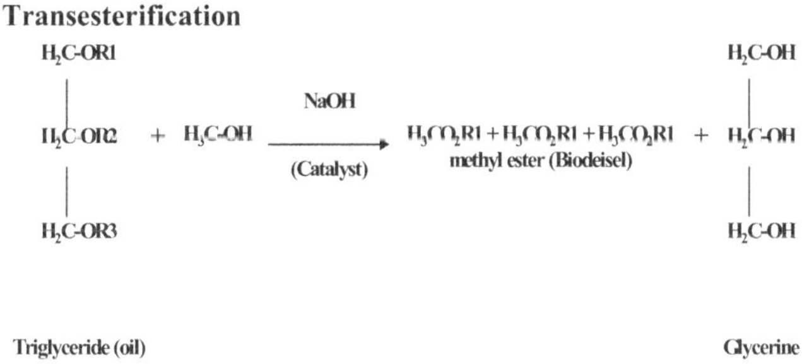

......,,...(7.11)

The raw manure input into the BEST system was 40,000 tons of pig manure from 257,000 finishing hogs producing about 58 million liters (ML) of biodiesel per year. This system with a daily flow of 6,500 m^3 was designed to produce 26,000 liters of methanol/d. With a biomethanol:oil ratio of 1:6, about 151,000 liters of biodiesel and 20,000 liters of glycerin were daily produced.

Assuming that the wastewater flow is only from the pig farms and the wastewater discharge from each pig is (6,500 × 1,000 liters/d)/257,000 pigs or 25 liters of wastewater/day/pig, this indicates that a large amount of treated effluent was recycled back as flush water. Since the gas production was 35,000 m^3/d, CH_4 production will be 23,000 m^3/d assuming 65% of the total gas was CH_4. Therefore, the biogas production /head/day was 0.136 m^3 biogas/head/d and the biomethanol production/head was 0.1 L/head/d or 0.6 L biodiesel/head/d

In Table 7.4, Equations (7.8) and (7.9) are the key chemical reactions that take place in the steam/methane reforming process. Equation (7.8) is endothermic, which means the desirable forward reaction is favored at high temperature and low pressure (~1600°F and ~200 psig). The water gas shift reaction, Equation (7.9), is exothermic. Carbon dioxide shifts the equilibrium towards the formation of the desired end product, synthesis gas, which is also known as syngas. It should be noted that commercial facilities producing methanol from natural gas need to buy carbon dioxide to be used as a reactant to force the equilibrium towards the formation of syngas. Equation (7.10) shows the formation of methanol (or biomethanol, as it is called, since it is produced from an organic feedstock of manure).

Equation (7.11) shows the transesterfication reaction. In this example, sodium hydroxide is used as a catalyst. The reaction occurs below 100°C (212°F) and about atmospheric pressure. As shown, the end products are glycerin and methyl esters (biodiesel), which are separated through distillation. R1, R2 and R3 represent alkyl radicals of different oils and fats that form the various methyl esters.

7.4 KOREAN PRACTICES WITH LIMITED LAND

Korea has a small area of 10×10^6 ha with 47 million people and about 10 million pigs resulting in 470 people/km^2 and 100 pigs/km^2 or 500 pigs/km^2 of agricultural area. On the average each farm has 700 pigs. The Korean Ministry of Environment (MOE) allows for the discharge of treated piggery effluents into the water environment. The effluent discharge regulation is very stringent particularly for centralized plants as shown in Table 1.2. Besides the tight control of traditional pollutants (BOD and SS), low values of COD, nitrogen and phosphorus are required. These restrictions have lead to the development of unique liquid treatment systems.

There are two separate systems for handling piggery wastewaters. The first is to recycle the total waste for agricultural use and the other involves the reuse of the waste along with liquid treatment systems prior to discharge into the water body. The centralized plants have been built to treat collected wastewater from the smaller farms for economical and technical reasons. Needless to say, the larger farms have their own treatment facilities.

The Korean government has already invested about 1.2 billion USD in the development of these systems. About 80 % of the total investment was provided for the construction of reuse plants of animal waste such as composting and bioliquid fertilization plants, and the remaining 20% was used for the construction of centralized wastewater treatment plants. The results of the investment for reuse was not really successful, because the land area was too limited. Therefore, the influent wastewater concentrations have been increased greatly because the solids are not separated properly for reuse.

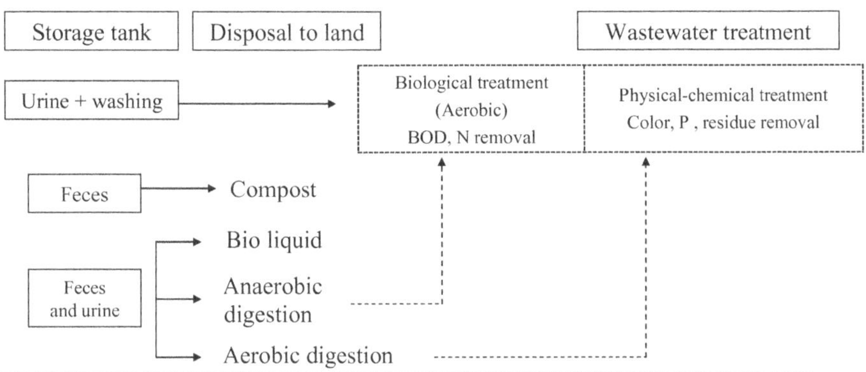

Removal efficiency	Biological treatment of whole waste*	Biological treatment of liquid only**	Chemical treatment	Membrane
COD (%)	< 80	> 95	99	99.5
N (%)	< 80	95	97	98
P (%)	-	70	99	99.5

*Biological treatment of whole liquid waste mixed with feces

** Biological treatment of liquid waste with previously separated feces

Figure 7.7 General treatment schemes of piggery waste in Korea (Choi, 2005).

Figure 7.7 presents a general scheme on how piggery wastes are handled in Korea. Wastewater is normally treated aerobically to comply with the effluent nitrogen requirement. Because anaerobic treatment removes carbon which is required for denitrification after aerobic treatment, it has not been used. The aerobically treated effluent has to be polished by physical and chemical processes to comply with the final effluent discharge regulation as shown in Table 1.2.

7.4.1 Waste Recycle Systems for Animal/Agriculture Uses

Waste recycle systems for animal/agriculture uses include composting, bioliquid fertilization, water reuse and nutrient recycling, which were previously discussed in sections 4.3, 4.4.2, 4.5 and 4.8, respectively.

(1) Composting

Feces and sawdust are mixed for 15 to 25 days to facilitate the bioremoval of easily degradable organics and an additional 20 to 30 days to stabilize the wastes prior to the land application. The optimum water content is reported to be 65%. The remaining liquid can be used for bioliquid fertilization or treated for final discharge either by individual (decentralized) plants or centralized plants.

(2) Bioliquid Fertilization

The minimum storage period for the production of a bioliquid from anaerobic digestion which is suitable for use as a fertilizer is 4 to 6 months prior to its application to the land. However, when air is applied (aerobic digestion), organics are removed after 10 days aeration and the supernatant can be used to clean pig houses or it can be stored for later application to agricultural land. In order to utilize this technique, the use of food additives has been recommended to minimize odor problems in feces.

(3)Water Reuse and Nutrient Recycling

Typically, the water quality from reuse plants could not easily meet the effluent regulatory requirements. However, the applications of membrane technologies to the existing plants allowed compliance with the effluent requirements. When reverse osmosis is used, the effluent can be reused and the nutrient in the condensate (or reject) can be recycled as fertilizer. The contents of N, P and K were 1,650, 382 and 2,059 mg/L respectively, as shown in Table 6.12. It was shown that the concentrate applied to arable soil with a recommended nitrogen application rate of 110 kg N/ha/yr yielded 6 tons/ha of rice with husk. This is similar to the yield with chemical fertilizer application as described earlier. Also, the leached nitrate was similar to that leached when chemical fertilizers were applied. In addition, the total available and labile phosphorus were similar to that obtained when conventional NPK fertilizer was used.

7.4.2 Wastewater Characteristics

Wastewater characteristics of farms vary depending on the ambient temperature, what kinds of feed are used, how solids are separated and how much washing water is used. Also, the feed type varies with the age of the pigs, but a kilogram of feed generally contains about 3,400 kcal of energy with 3,500 mg chloride, 0.4 to 0.7 % P_2O_5, 40 to 100 mg Fe and 50 to 100 mg Zn as recommended by NRC (1998).

Wastewater production including the feces is about 5 to 8 L/head/d depending on climate as mentioned earlier. Feces have a very high COD with low nitrogen, while urine has less COD with high nitrogen This will results in various wastewater strengths depending upon how the solids are separated. Korean practices have shown that influent concentrations of 50 g/L COD, 5 g/L N and 600 mg/L P have been considered as a typical for the design of centralized plants. However, there were some plants with about 10 g/L COD and about 3.5 g/L N.

7.4.3 Wastewater Treatment Systems

Even with the encouragement for animal fertilizer use, the amount of excess animal wastes have been increasing since farm sizes have been getting bigger with limited agricultural area. As of 2004, there were 75 centralized community plants treating a total of 12,510 m^3/d of piggery waste. Among them, 41 plants were operating with a total of 9,745 m^3/d and the remaining 34 plants, which were to produce 2,765 m^3/d, were under construction. As indicated earlier, a few plants utilize the anaerobic process followed by an aerobic process, but most plants prefer to use the aerobic process followed by a physical and chemical polishing stage to comply with the effluent requirements. There are some joint plants treating piggery and municipal sewage which do not require a polishing stage. In general, piggery wastewater treatment plants consist of solid/liquid separation or other pretreatment processes, main treatment and post-treatment polishing stages (see Chapter 6).

(1) Pretreatment

Drum screens are mostly used to remove larger debris and solids followed by centrifugation to remove smaller solids. Struvite precipitation, and ammonia stripping are not used to control the COD/N ratio which is a critical parameter for a successful biological nutrient removal process. Normally flow equalization is used with a 5 day HRT after the solid separation process.

(2) Anaerobic Digestion

This is an excellent process to remove organics, but it can not remove nitrogen. This results in a lower COD/N ratios in the aerobic stage. One of the existing Korean plants removed 68 to 76% BOD with 0.4 to 1.2 kg BOD/m^3/d. The high pH of the piggery waste resulted in a high methane content and methane yield of 0.29 to 0.32 m^3 CH_4 produced/kg COD removed.

(3) Aerobic Process

This process is applicable to a weaker wastewater (in comparison to the anaerobic digestion process) since it requires aeration, but it has the merit that it can remove both organics and nitrogen. Generally, it is designed for a BOD load of 0.7 kg/m^3/d and a 0.25 kg N/m^3/d load for nitrification and denitrification. If only nitrite nitrification is sought, the tank volume requirement can be reduced and about 40% of the carbon can be saved. However, special care must be taken since nitrification can be inhibited by the heat released from carbon oxidation during the summer months when the influent strength exceeds 30 g/L COD with an organic load of 0.5 kg COD/m^3/d.

In order to reduce the influent COD concentration to the aerobic process, an anaerobic digester can be used as a pretreatment process. However, in order not to avoid reduced denitrification efficiencies with limited carbon, it is suggested that a part of flow be bypassed around the anaerobic digester using the Modified Ludzak and Ettinger (MLE) scheme (Kim *et al.,* 2004). About 40 to 80% bypass flow was recommended depending on the desired reactor temperature.

(4) Polishing Stage (Post Treatment)

At present, biological oxidation is the major process for the piggery waste treatment. However, there are still some nonbiodegradable organics remaining in the effluent such as color, which requires a subsequent polishing stage. If the operating MLSS is high, a mechanical thickening process is utilized with a chemical aid for solids separation. Membrane separation is also used for this purpose. AOP and carbon filtration are used for color and residual COD removal.

7.4.4 Evaluation of Three Full-scale Plants

The nitrogen removal performances of three full-scale plants with different organic and nitrogen loads and with flow ranges of 95 to 130 m^3/d were compared (Choi *et al.,* 2005). Plants 1 and 2 were modifications of anoxic-aerobic systems, and Plant 3 was an anaerobic and anoxic-aerobic system as shown in Figure 7.8 and Table 7.5. Plant 1 had relatively higher organic and nitrogen loads resulting in higher COD and NH_4-N levels in the effluent. Plant 2 had higher influent COD levels resulting in higher operating temperatures. Although Plant 3 was operated with higher influent COD/TKN ratios, an additional external carbon source was required to polish the final effluent to facilitate nitrogen removal. Influent COD in Plant 3 was used in the anaerobic/anoxic reactors for both methane production and denitrification. Based on various mass balances, including caloric, COD, and alkalinity, the key elements for the successful nitrogen removal from the piggery wastewater are reactor temperature, influent COD/TKN and alkalinity/TKN ratios as summarized in Table 7.5.

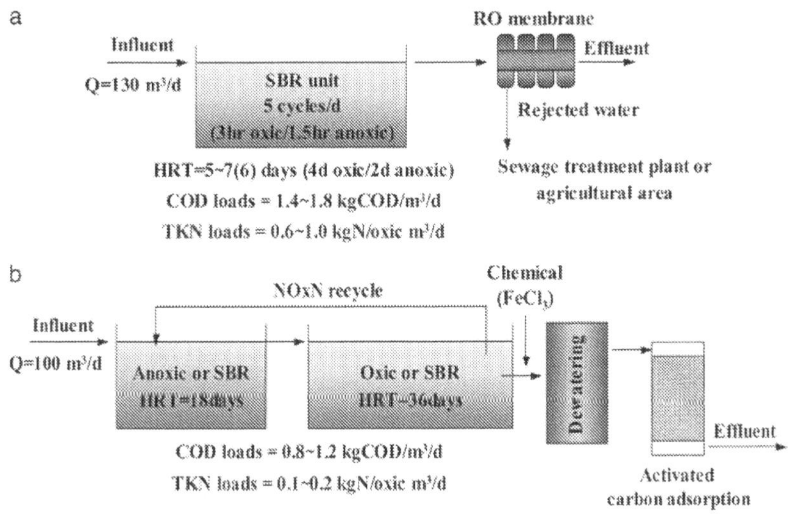

Figure 7.8 Schematics of full-scale biological treatment plants evaluated by Choi *et* al (2005). Pretreatment, such as degritting, screening and flow equalization, is not shown for Plants 1 and 2 and the polishing stage for Plant 3 is not shown.. (a) Plant 1 - Sequencing-batch reactor with RO polishing; (b) Plant 2 - Sequencing-batch reactor and modified Ludzack-Ettinger process with activated carbon polishing;

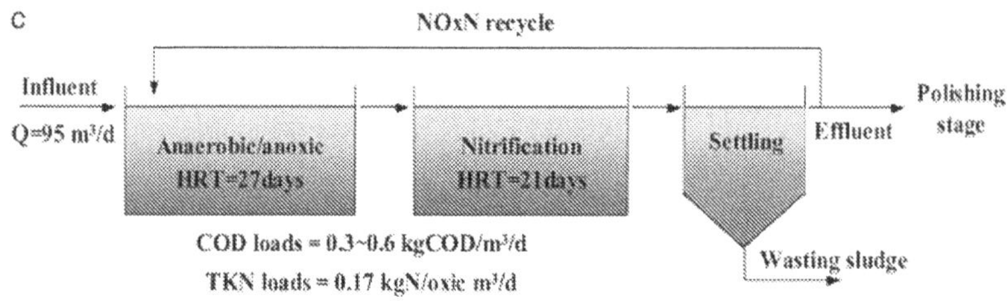

Figure 7.8 (continued) Schematics of full-scale biological treatment plants evaluated by Choi et al (2005). Pretreatment, such as degritting, screening and flow equalization, is not shown for Plants 1 and 2 and the polishing stage for Plant 3 is not shown. (c) Plant 3-Anaerobic/anoxic and aerobic processes.

Table 7.5 Performance summary for three full-scale bioreactors plants (Choi *et al.*, 2005).

	Plant 1 Influent (g/L)	Plant 1 Effluent(1) (mg/L)	Plant 2 Influent (g/L)	Plant 2 Effluent(2) (mg/L)	Plant 3 Influent (g/L)	Plant 3 Effluent(3) (mg/L)
Temperature(°C)	-	25-41	-	36-47	-	30-35
pH	8.8	8.0	8.3	8.3	8.4	6.6
Alkalinity	10.0	2,300	NA	NA	6.2	90
TCOD	9.80	2,908	45.32	NA	25	1,219
SCOD	6.73	1,195	29.46	1,230	13.6	1,083
TKN	2.87	487	6.68	NA	3.5	17
Soluble NBD N	-	14	-	47	-	16
NH_4-N	1.72	150	3.47	40	2.97	13
NO_2-N	0	160	0	47	0	0
NO_3-N	0	85	0	15	0	458
TCOD/TKN	3.4	-	6.8	-	7.1	-
Alk/TKN	3.5	-	NA	-	1.8	-
COD loading rates (kgCOD/m^3/d)	1.4-1.8		0.6-1.2		0.3-0.6	
TKN loading rates (kgTKN/m^3/d)	0.6-1.0		0.1-0.2		0.17	
Oxic/total (m^3/m^3)	0.6		0.7		0.4	

(1) Plant 1: Effluent concentrations excluding membrane separation,

(2) Plant 2: Effluent concentrations before dewatering and excluding summer months

(3) Plant 3: Effluent concentrations without the final denitrification reactor

Nitrogen removal in piggery wastewater is greatly affected by the operating temperature of the processes. Nitrogen mass balances were performed for the operating temperature ranges in the study (Choi *et al.,* 2005). Figure 7.9 summarizes how the nitrogen was removed at different temperatures from Plant 2. Basically, biological nitrification and denitrification dominated at temperatures lower than 35°C. Above 35°C, ammonia stripping becomes a dominant removal mechanism because the nitrifying activity decreased at higher temperatures. Biological nitrogen removal was greatly reduced at temperatures higher than 40°C. Predicted and measured operating temperatures are shown in Figure 7.10.

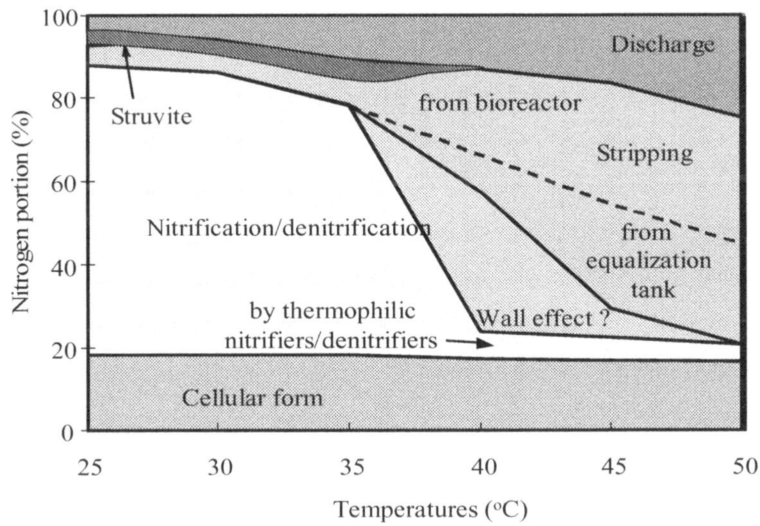

Figure 7.9 Estimated nitrogen removal efficiencies by mass balances for different temperatures. Ammonia loss by stripping from the bioreactor and equalization tank increased as temperature increased. A biofilm growth of nitrifiers and denitrifers on tank walls and bottom, which was cooled by the surrounding soil, seemed to remove some nitrogen, even at temperatures higher than 35°C. Struvite formation could contribute to nitrogen removal throughout the temperature ranges.

Nitrification, without alkalinity addition, was well developed at the three plants until the reactor temperature reached 35°C. The ammonium uptake rate (AUR) and NO_x-N uptake rate (NUR) increased as the NO_2-N/NO_3-N ratios in the reactor increased. Nitrite accumulation was observed when the reactors operated at high nitrogen and pH. The AUR ranged from 7 to 24 mg NH_4-N/L/h at 20 to 35°C, depending on the NO_2-N/NO_3-N ratios in the aerobic reactor. The denitrification rates also varied in the ranges of 4.2 to 13.5 mgNO_x-N/L/hr at 20 to 35°C. Based on these observations, it can be concluded that nitrite nitrification has beneficial effects on nitrogen removal.

The Plant 3 anaerobic-anoxic reactor utilized a combined UASB and fixed-media system configuration which produced approximately 280 m^3 gas/d. The gas contained approximately 66% N_2 and approximately 33% CH_4. The carbon dioxide content in the gas was less than 1%.

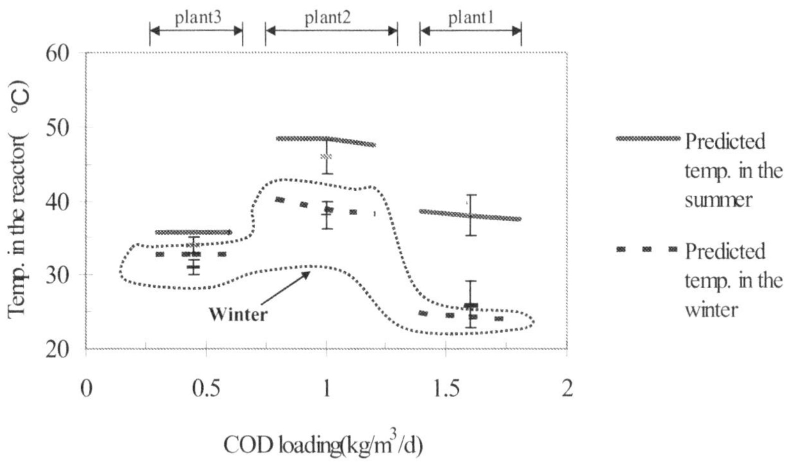

Figure 7.10 Predicted and measured temperatures at Plants 1, 2 and 3. The figure is based on the following assumptions: the heat transfer coefficient in the reactor wall is 20 kcal/m^2/°C/d; the heat produced from COD removed is 3,500 kcal/kg $COD_{removed}$; the ambient temperatures were 10°C in winter and 25°C in summer; and the air temperature from the blower were 35°C in winter and 57°C in summer.

7.4.5 PJ Plant

The PJ plant has a treatment capacity of 110 m^3/d with 50 m^3/d of kitchen waste and 60 m^3/d of piggery waste. Originally this plant was built to treat 30 m^3/d of kitchen waste, but, later an 80 m^3/d plant was constructed to treat 20 m^3/d of kitchen waste and 60 m^3/d piggery waste as shown in flow diagram (Figure 7.11). The kitchen waste was segregated to remove nonbiodegradable wastes and then ground prior its introduction into an anaerobic digester. The animal waste was screened prior its addition to another anaerobic digester. After digestion, both digester effluents were combined for solids separation using dissolved air flotation (DAF) followed by a biological nutrient removal process involving anoxic and oxic reactors (MLE scheme). The effluent from the BNR system was further treated in a series of membrane processes including MF, UF and RO. The designed influent flow rate was 224 m^3/d including the reject flow from sludge processing and considering the seasonal variations in water demand for the animals. The gas production was about 40 m^3 (65% CH_4) /m^3 piggery waste and 60 m^3 (65%) /m^3 kitchen waste. On one day, the total gas producted was 4,920 m^3 with 67% CH_4, and about 300 KWe was produced with 90 m^3/d hot water at 90°C which was recovered by cooling the gas engine. The hot water was used for heating the digester and the spent water, which was about 25 m^3/d at 70°C, was recycled back to this plant as process water. Moisture in the biogas was removed before gas storage in a 450 m^3 tank. The gas was pressurized prior to application to the gas engine which can produce about 500 KWH. The operating results for this plant are given in Table 6.8.

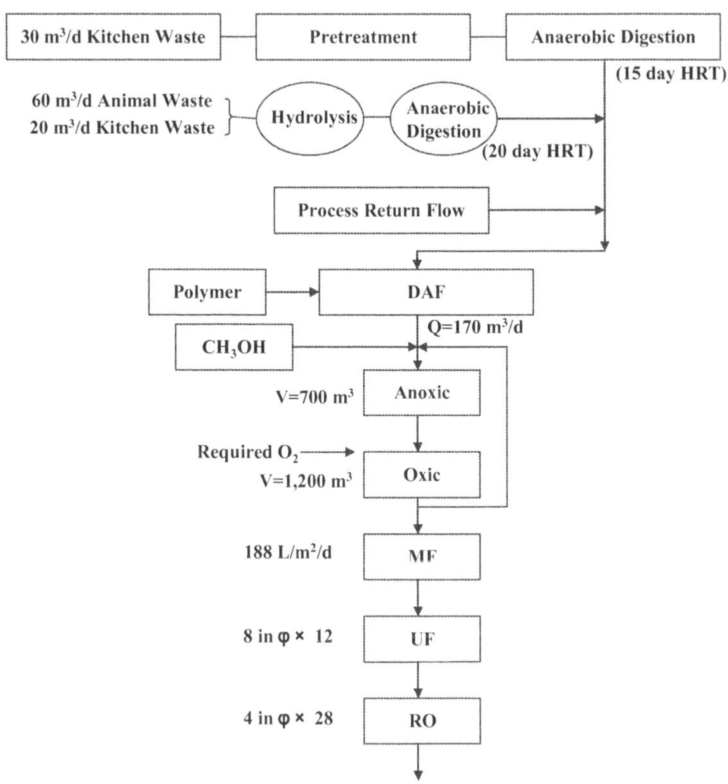

Figure 7.11 Schematic flow diagram of the PJ plant.

7.4.6 Integrated Biogas Energy System (IBES)

The IBES system (Chae *et al.,* 2001) shown in Figure 7.12 includes an anaerobic digester and an aerobic biofilm system followed by a reverse osmosis process. The biogas generated from the anaerobic digester was used for electrical generation. The 200 m^3 anaerobic digester was operated at 20 day SRT with a loading of 0.5 to 3 kg $VS/m^3/d$. To avoid ammonium toxicity in the anaerobic digester, the pH was maintained less than 7.5 and gas mixing was provided inside the digester. Gas storage was provided for about 25 to 33% of the gas production per day. The gas was used for electrical generation. Gas generation was about 0.52 to 0.72 m^3/kg VS applied which is a little higher than the typical production rate of 0.3 to 0.4 m^3/kg VS applied. The CH_4 content was 68 to 73%. The dual fuel engine generator with a capacity of 37 KWe was operated about 12 hours a day to avoid overload. Ferric chloride was added into the sludge recirculation line to prevent sulfide gas formation and potential corrosion problems. The VS reduction efficiency was about 50% and the gas generation was 80 to 140 m^3/day.

The digested liquor was intended to be used for agricultural application. However, further treatment was necessary when this use was not allowed. The treatment included aerobic biofilm treatment followed by reverse osmosis polishing.

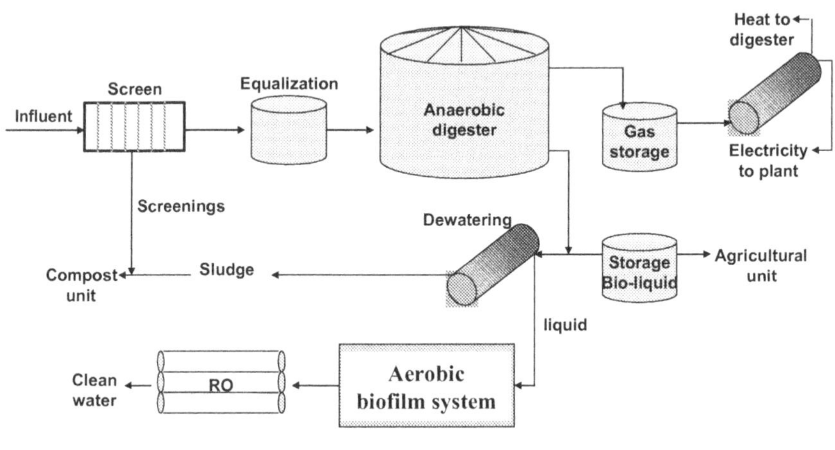

Figure 7.12 Integrated biogas energy system (IBES) developed in Korea.

7.4.7 Pig Apartment

A pig farm constructed two pig apartments, one for the mother and baby pigs, and the other for the adult pigs. One such apartment is shown in Figure 7.13. It can save land and its owner says it is easy to move pigs using elevators. Two separate buildings are convenient to provide different room temperatures. The baby pigs need warmer temperatures, while the adult need cooler temperature. Also, the separate apartments help to block disease transmission from one building to another.

The total floor area for both apartments is about 3,600 m^2 for 5,000 to 7,000 pigs. The wastes are collected through slotted floors and are transferred to an underground storage tank located adjacent to the buildings. The tank storage capacity is about 800 m^3, which is about a 20 day HRT. The settled solids are pumped out of the storage tank and screened to produce a compostable residual. The liquid from the storage tank is pumped to an aerated storage tank with a capacity of about 480 m^3. Aeration produces a liquid fertilizer. All wastes from the farm are recycled for agriculture use. The tank storage was covered and the area was used for composting.

Figure 7.13 Pig apartments. There are two identical three-story apartments, one for the mother and baby pigs and the other for adult pigs. This picture shows only one apartment.

(a) Aerobic digester for bioliquid

(b) Composting area

Figure 7.14 Aerobic digester for bioliquid and composting area for waste recycling. There is a waste storage tank under the compost area.

7.4.8 Pig house simulated to nature

Pigs love digging holes and playing around. A farmer constructed a pig house with about 1 m depth of compost floor mixed with saw dust and soil (10 to 1 ratio). Pigs are growing on this floor. The pig house is designed to receive ample sunlight and ventilated well and there is no odor and waste management problems. The wastes from pigs are absorbed into the compost floor, which requires only 1 time of mixing from bottom to top layers a year. It was found that pigs are healthy and require less medicine for 3 years operation (KBS 1, 2007).

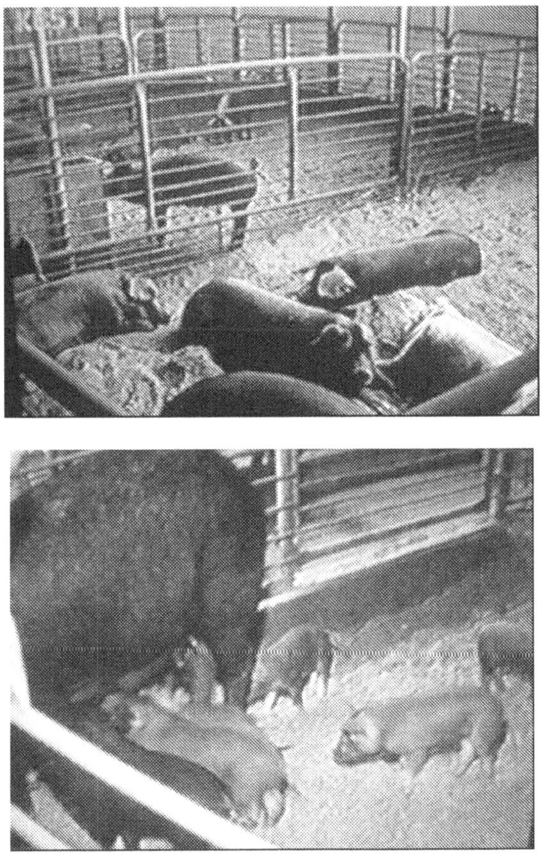

Figure 7.15 A pig house with compost floor simulated to nature. Pigs are growing in more space than a caged concrete or iron floor. The wastes produced from pigs are absorbed into compost floor.

7.5 TROPICAL EXPERIENCES

7.5.1 Singapore Experience

Although pig farming in Singapore has ceased because of the competing need for housing and industrial land, the information contained in a previous study on pig waste management in Singapore (Taiganides, 1992) would help to solve problems in other tropical areas.

The wastewater discharge in Singapore during the 1980's was 20 L/pig/day, which is about four times the wastewater discharged during cold winter operation in Korea. Besides weather differences, the higher discharge could be the result of an uncontrolled water supply for the pigs. Today, water is only supplied for drinking by specially designed faucets, which are not used for bathing or cooling down the body temperature of the pigs. The daily unit discharge from each pig in the Singapore study was 130 g BOD, 450 g COD, 30 g N, 10 g P and 5 g K.

The Singapore initial facility was a total recycling system with algal harvesting as shown in Figure 7.15. The revamped treatment systems consisted of anaerobic digestion with methane generation in an anaerobic lagoon followed by an oxidation ditch system for recycling water for flushing as shown in Figure 7.16.

Figure 7.15 is the flow scheme with algae harvesting. An influent with 4,260 mg/L BOD and 12,160 mg/L TSS was introduced to a settling tank and its effluent went to a 30 cm deep algal pond which had an organic loading of 200 to 400 kg BOD/ha/d and a detention time of 4 to 6 d. Algal harvesting was accomplished using centrifugation, dissolved air flotation and bioflocculation with chitosan. There were two algal ponds installed. Each demonstration pond was 7 m wide, 90 m long and had a total surface area of 1,250 m^2. Each pond consisted of a single loop (or race track configuration) with a paddle wheel mixer and a depressed section for the removal of settled solids. Gross biomass productivities expressed in TSS were 16.6 for 8 day HRT, and 5.7 $g/m^2/d$ for 16 day HRT, respectively. A HRT of 4 to 8 days was the best. The dewatered biomass was pasteurized with heat recovered from the anaerobic digester. The primary settling tank was designed with 1.9 hr HRT and the anaerobic digester was designed with a one to 2.2 kg $COD/m^3/d$ loading rate and a 12 to 30 day HRT. The digester was operated at 25 to 29°C without external heating. The gas production was 0.32 to 0.51 m^3/kg COD or 0.26 to 0.54 m^3/kg VS with an influent of 20.8 to 34 g/L COD. About 71% of the gas produced was CH_4. Generally, about 0.22 to 0.26 m^3 gas/day was produced from each pig.

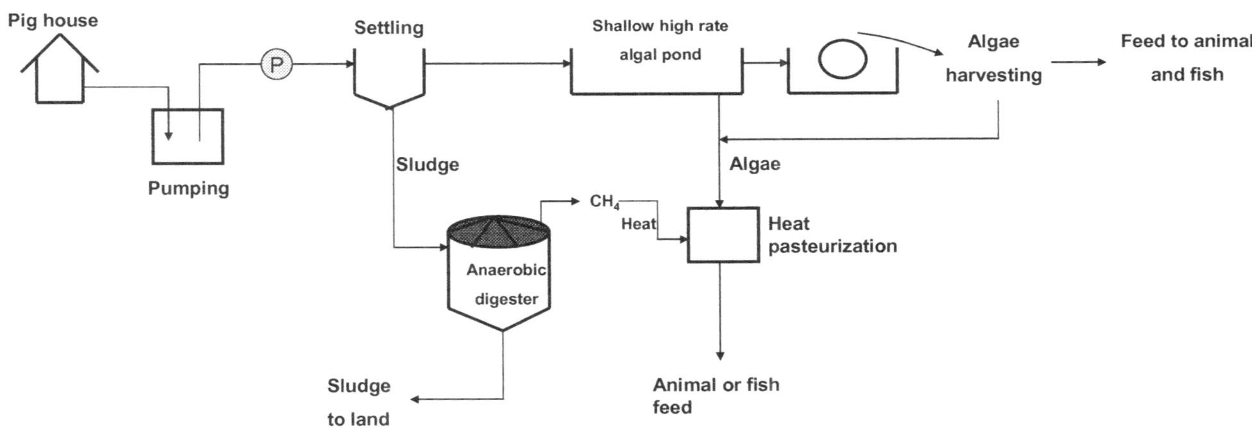

Figure 7.15 Singapore studied tctal recycling system in 1985(Taiganides, 1992).

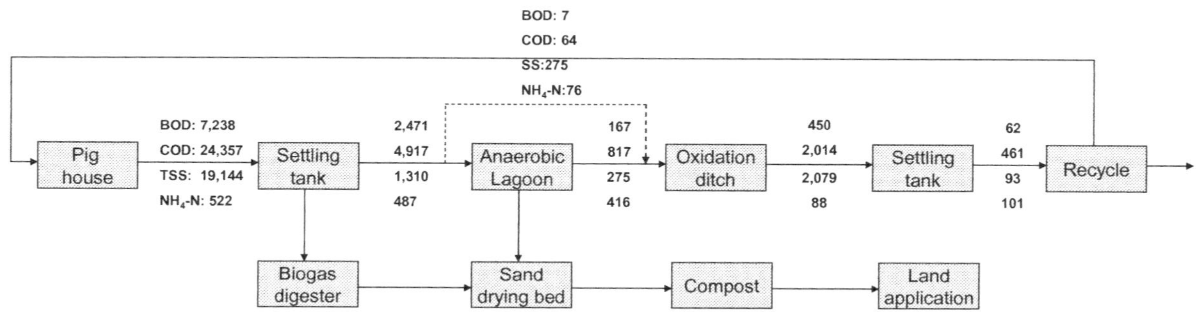

Figure 7.16 Singapore total recycling system with changes of organic and NH_4-N concentrations (mg/L). The anaerobic lagoon shown in this figure was converted later into an aerobic lagoon by the installation of surface aerators.

In addition to algae harvesting, the possibility of using wasted activated sludge from an oxidation ditch as feed material for the pigs was also studied. A greater incidence of pneumonia was observed in pigs whose diets contained the bacterial biomass. However, no deaths were recorded.

Depending on the applicable discharge regulations three different treatment systems were purposed and included:

(1) Solids stabilization treatment with 5 m deep anaerobic lagoon, sand fil ter and sludge composting system for an effluent BOD of less than 50 0 mg/L (90% removal). The lagoon had 5 m depth with 37 day HRT o r 0.09 kg BOD/m^3/d.

(2) The most practicable treatment system was achieved by adding surfac e aerators to the lagoon followed by a secondary settling tank aided by chemical coagulation. The effluent quality to be achieved was 250 mg /L BOD plus solids stabilization, and

(3) A best applicable treatment was designed to demonstrate unit operatio ns that would produce BOD and TSS less than 50 mg/L using an oxid ation ditch and recover energy capability by stabilizing the solids usin g a fully controlled anaerobic digester.

Figure 7.16 shows the change of organic and nitrogen concentrations with the best applicable treatment system with the recycled water for flushing. The settled wastewater feed was split into aerobic lagoon and oxidation ditch to make a uniform loading. The final effluent was recycled back to pig house.

7.5.2 Malaysian Experience

The Malaysians reported in May 2002 that they used a flow rate of 30 L/d/pig for design purposes when the author visited two of their treatment plants. Both plants were operated to comply with an effluent BOD requirement of less than 50 mg/L. The influent was screened prior to its introduction to the treatment systems.

Plant A provided anaerobic lagoon treatment with about a 20 day HRT followed by aeration basin with a 2.5 day HRT and that followed by a settling tank with 2.5 day HRT. Plant B had a similar treatment scheme except it had an anaerobic lagoon with 12 day HRT followed by a sequencing batch reactor (SBR) with 1 day HRT operating on a 1 hour anaerobic and a 1 hour aerobic cyclic operation. The sludge, which was produced, was filtered with a cloth filter and then dried on a sand bed. The dried sludge was used to fertilize nearby golf courses.

REFERENCES

Banks C J, Salter A M and Chesshire M (2007) Potential of anaerobic digestion for mitigation of greenhouse gas emissions and production of renewable energy from agriculture: Barriers and incentives to widespread adoption in Europe, *Wat. Sci. Technol.* **55**(10), 165-173.

Bull L S (2005) Innovative sustainable systems utilizing economical solutions (ISSUES) A 3 component program, *Final report for the NC Attorney General-Smithfield Food/Premium Standard Farms/Frontline Farmers Agreement*, NC State University.

Bull L S and Cook M (2004) Orbit/HSAD, *EST on campus report*, NC State University, Raleigh, NC

Chae K J, Lim S K, Choi K H, Park W K, Lim D K (2001) Integrated biogas energy system, *J of KOWREC*, **9**(3), 19.

Choi E (2005) Piggery waste treatment systems in Korea: Current practices, *Presented at A&WMA and WEF international conference on animal agriculture and processing*, St. Louis, MO.

Choi E, Kim D, Eum Y, Yun Z and Min K S (2005) Full scale experience with nitrogen removal from piggery waste, *Wat. Env. Res.* **77**(4), 381-388.

Dugba P (2003) Conversion of swine manure into bio-diesel using anaerobic digesters, *Proc of anaerobic digester technology applications in animal agriculture – a national summit*, June 2-4, Raleigh, NC ,WEF.

Hansen K H, Angelidaki I and Ahring B K (1999) Improving thermophilic anaerobic digestion of swine manure, *Wat. Res.* **33**(8), 1805-1813.

Humenik F (2004) Introduction in *Animal waste containment in lagoons,* edited by Reddi, L.N., 6, *ASCE*.

Humenik F J, Rice J M, Baird C L and Koelsch R (2004) Environmentally superior technologies for swine waste management, *Wat. Sci. Technol.* **49**(5-6), 15-22.

KBS 1 (2007) Pigs is cages, Environmental Special broadcasted by Korean Broadcasting System 1, 23 May 2007

Kim D, Choi E, Yun Z and Kim S (2004) Nitrogen removal from piggery waste with anaerobic pretreatment, *Wat. Sci. Technol.* **49**(5-6), 165-171.

Kottner M (2004) The role of biogas in German farm business development, *bioexell European biogas conference –biogas in society*. Enniskillen, Co. Fermanagh, Northern Ireland 21-23 October.

Lens P, Westermann P, Haberbauer M and Moreno A (2006) Biofuels and fuel cells: renewable energy from biomass fermentation, *Water 21*, February, 41-48.

Miner J R, Humenik F J, Rice J M and Rashash D (2003) Development of a biologically active, permeable foam material for livestock producers concerned with odor control and ammonia, *IWA specialty conference on strong nitrogenous and agro-wastewater*, June 11-13, Seoul, Korea.

National Research Council (1998) *Nutrient requirement of swine*, tenth revised edition, subcommittee on animal nutrition, Board on Agriculture, The national Academy of Sciences.

Taiganides E P (1992) *Pig waste management and recycling*, IDRC, Ottawa, Ont. Canada.

Van herle J, Membrez Y And Bucheli O (2004) Biogas as a fuel source for SOFC cogenerators. *Journal of Power sources,* 127, 300-312.

Wiese J and Haeck M (2006) Instrumentation, control and automation for full-scale manure based biogas systems, *Wat. Sci. Technol.* **54**(9), 1-8.

Wiltsee G and Emerson H (2003) Clean and reliable power and heat from digester gas, *Anaerobic digester technology applications in animal agriculture- A national summit,* Raleigh, NC, WEF.

Index

IWA Publishing's authorised EU representative for General Product Safety Regulations is Diane D'Arras, 15 rue Duret, 75116 Paris, France, e-mail: safety@iwap.co.uk.

Printed and bound by CPI Group (UK) Ltd, Croydon, CR0 4YY
16/06/2026
02140382-0001